OIL WELL DRILLING TECHNOLOGY

OIL WELL DRILLING TECHNOLOGY

Deepak Sharma

RANDOM

RANDOM PUBLICATIONS

NEW DELHI - 110 002 (INDIA)

Oil Well Drilling Technology

ISBN 978-93-51116-68-4

Published in 2015 in India by

RANDOM PUBLICATIONS

4376-A/4B, Gali Murari Lal, Ansari Road
New Delhi-110 002
Phone: +9111-43580356, 23289044
E-mail: randomexports@gmail.com; sales@randompublications.com; info@randompublications.com

Reprinted 2026

Type Setting by: Friends Media, Delhi-110089

Preface

An oil well is a boring into the Earth that is designed to bring petroleum oil hydrocarbons to the surface. Usually some natural gas is produced along with the oil. A well that is designed to produce mainly or only gas may be termed a gas well. The well is created by drilling a hole 12 cm to 1 metre (5 in to 40 in) in diameter into the earth with a drilling rig that rotates a drill string with a bit attached. After the hole is drilled, sections of steel pipe (casing), slightly smaller in diameter than the borehole, are placed in the hole. Cement may be placed between the outside of the casing and the borehole. The casing provides structural integrity to the newly drilled wellbore, in addition to isolating potentially dangerous high pressure zones from each other and from the surface. With these zones safely isolated and the formation protected by the casing, the well can be drilled deeper with a smaller bit, and also cased with a smaller size casing. Modern wells often have two to five sets of subsequently smaller hole sizes drilled inside one another, each cemented with casing.

This process is all facilitated by a drilling rig which contains all necessary equipment to circulate the drilling fluid, hoist and turn the pipe, control downhole, remove cuttings from the drilling fluid, and generate on-site power for these operations. After drilling and casing the well, it must be 'completed'. Completion is the process in which the well is enabled to produce oil or gas. In a cased-hole completion, small holes called perforations are made in the portion of the casing which passed through the production zone, to provide a path for the oil to flow from the surrounding rock into the production tubing. In open hole completion, often 'sand screens' or a 'gravel pack' is installed in the last drilled, uncased reservoir section. These maintain structural integrity of the wellbore in the absence of casing, while still allowing flow from the reservoir into the wellbore. Screens also control the migration of formation sands into production tubulars and surface equipment, which can cause washouts and other problems, particularly from unconsolidated sand formations of offshore fields. After a flow path is made, acids and fracturing fluids may be pumped into the well to fracture, clean, or otherwise prepare and stimulate the reservoir rock to optimally produce hydrocarbons into the wellbore. Finally, the area above the reservoir section of the well is packed off inside the casing, and connected to the surface via a smaller diameter pipe called tubing. This arrangement provides a redundant barrier to leaks of hydrocarbons as well as allowing damaged sections to be replaced. Also, the smaller cross-sectional area of the tubing produces reservoir fluids at an increased velocity in order to minimize liquid fallback that would create additional back pressure, and shields the casing from corrosive well fluids. The production stage is the most important stage of a well's life, when the oil and gas are produced. By this time, the oil rigs and workover rigs used to drill and complete the well have moved off the wellbore, and the top is usually outfitted with a collection of valves called a Christmas tree or production tree. These valves regulate pressures, control flows, and allow access to the wellbore in case further completion work is needed. From the outlet valve of the production tree, the flow can be connected to a distribution network of pipelines and tanks to supply the product to refineries, natural gas compressor stations, or oil export terminals. Workovers are often necessary in older wells, which may need smaller diameter tubing, scale or paraffin removal, acid matrix jobs, or completing new zones of interest in a shallower reservoir. Depending on the type of lift system and wellhead a rod rig or flushby can be used to change a pump without pulling the tubing. New drilling technologies including satellites, global positioning systems, remote sensing devices, and 3-D and 4-D seismic technologies make it possible to discover oil reserves while drilling fewer wells.

The present book is an effort to familiarize the students and professionals with the latest development in the field.

I thank all members of my team who have helped in the preparation of the book. My special thanks go to "Random Publications" who have published the book.

— Deepak Sharma

Contents

Chapter 1

Introduction

Blowout (Well Drilling)

A blowout is the uncontrolled release of crude oil and/or natural gas from an oil well or gas well after pressure control systems have failed. Prior to the advent of pressure control equipment in the 1920s, the uncontrolled release of oil and gas from a well while drilling was common and was known as an oil gusher, gusher or wild well.

History

Figure: *The Lucas Gusher at Spindletop, Texas (1901)*

Gushers were an icon of oil exploration during the late 19th and early 20th centuries. During that era, the simple drilling techniques such as cable-tool drilling and the lack of blowout preventers meant that drillers could not control high-pressure reservoirs. When these high pressure zones were breached the hydrocarbon fluids would travel up the well at a high rate, forcing out the drill string and creating a gusher. A well which began as a gusher was said to have "blown in": for instance, the Lakeview Gusher *blew in* in 1910. These uncapped wells could produce large amounts of oil, often shooting 200 feet (60 m) or higher into the air.

A blowout primarily composed of natural gas was known as a *gas gusher*. Despite being symbols of new-found wealth, gushers were dangerous and wasteful. They killed workmen involved in drilling, destroyed equipment, and coated the landscape with thousands of barrels of oil; additionally, the explosive concussion released by the well when it pierces an oil/gas reservoir has been responsible for a number of oilmen losing their hearing entirely; standing too near to the drilling rig at the moment it drills into the oil reservoir is extremely hazardous. The impact on wildlife is very hard to quantify, but can only be estimated to be mild in the most optimistic models — realistically, the ecological impact is estimated by scientists across the ideological spectrum to be severe, profound, and lasting.

To complicate matters further, the free flowing oil was — and is — in danger of igniting. One dramatic account of a blowout and fire reads,

> *"With a roar like a hundred express trains racing across the countryside, the well blew out, spewing oil in all directions. The derrick simply evaporated. Casings wilted like lettuce out of water, as heavy machinery writhed and twisted into grotesque shapes in the blazing inferno."*

The development of rotary drilling techniques where the density of the drilling fluid is sufficient to overcome the downhole pressure of a newly penetrated zone meant that gushers became avoidable. If however the fluid density was not adequate or fluids were lost to the formation, then there was still a significant risk of a well blowout.

In 1924 the first successful blowout preventer was brought to market. The BOP valve affixed to the wellhead could be closed in the event of drilling into a high pressure zone, and the well fluids contained. Well control techniques could be used to regain control of the well. As the technology developed, blowout preventers became standard equipment, and gushers became a thing of the past.

In the modern petroleum industry, uncontrollable wells became known as blowouts and are comparatively rare. There has been a significant improvement in technology, well control techniques and personnel training that has helped to prevent them occurring. From 1976 to 1981, 21 blowout reports are available.

Notable Gushers

1. The Shaw Gusher in Oil Springs, Ontario, was North America's (and possibly the world's) first oil gusher. On January 16, 1862, it shot oil from over 60 metres (200 ft) below ground to above the treetops at a rate of 3,000 barrels (480 m^3) per day, triggering the oil boom in Lambton County.
2. Lucas Gusher at Spindletop in Beaumont, Texas in 1901 flowed at 100,000 barrels (16,000 m^3) per day at its peak, but soon slowed and was capped within nine days. The well tripled U.S. oil production overnight and marked the start of the Texas oil industry.
3. Masjed Soleiman, Iran in 1908 marked the first major oil strike recorded in the Middle East.
4. Dos Bocas'" in the State of Veracruz, Mexico, was a famous Mexican blowout that formed a large crate, and leaked oil from the main reservoir for many years, even after Pemex nationalized the Mexican oil industry in March 1938.
5. Lakeview Gusher on the Midway-Sunset Oil Field in Kern County, California of 1910 is believed to be the largest-ever U.S. gusher. At its peak, more than 100,000 barrels (16,000 m^3) of oil per day flowed out, reaching as high as 200 feet (60 m) in the air. It remained uncapped for 18 months, spilling over 9 million barrels (1,400,000 m^3) of oil, less than half of which was recovered.
6. A short-lived gusher at Alamitos #1 in Signal Hill, California in 1921 marked the discovery of the Long Beach Oil Field, one of the most productive oil fields in the world.
7. The Barroso 2 well in Cabimas, Venezuela in December 1922 flowed at around 100,000 barrels (16,000 m^3) per day for nine days, plus a large amount of natural gas.
8. Baba Gurgur near Kirkuk, Iraq, an oilfield known since antiquity, erupted at a rate of 95,000 barrels (15,100 m^3) a day in 1927.
9. The Wild Mary Sudik gusher in Oklahoma City, Oklahoma in 1930 flowed at a rate of 72,000 barrels (11,400 m^3) per day.

10. The Daisy Bradford gusher in 1930 marked the discovery of the East Texas Oil Field, the largest oilfield in the contiguous United States.
11. The largest known 'wildcat' oil gusher blew near Qom, Iran on August 26, 1956. The uncontrolled oil gushed to a height of 52 m (170 ft), at a rate of 120,000 barrels (19,000 m^3) per day. The gusher was closed after 90 days' work by Bagher Mostofi and Myron Kinley (USA).
12. One of the most troublesome gushers happened on June 23, 1985 at the well #37 at the Tengiz field in Atyrau, Kazakh SSR, Soviet Union, where the deep, 4209 metre well blew out and the 200-metres high gusher self-ignited two days later. Oil pressure up to 800 atm and high hydrogen sulfide content had led to the gusher being capped only on 27 July, 1986 when the well was closed by the shaped charge. The total volume of erupted material measured at 4.3 millions metric tons of oil, 1.7 bn m^3 of natural gas, and the burning gusher resulted in 890 tons of various mercaptans and more than 900000 tons of soot released into atmosphere.
13. The largest underwater blowout in U.S. history occurred on April 20, 2010, in the Gulf of Mexico at the Macondo Prospect oil field. The blowout caused the explosion of the Deepwater Horizon, a mobile offshore drilling platform owned by Transocean and under lease to BP at the time of the blowout. While the exact volume of oil spilled is unknown, as of June 3, 2010 (2010 -06-03), the United States Geological Survey (USGS) Flow Rate Technical Group has placed the estimate at between 35,000 to 60,000 barrels (5,600 to 9,500 m^3) of crude oil per day.

Cause of Blowouts

Reservoir Pressure

Petroleum or crude oil is a naturally occurring, flammable liquid consisting of a complex mixture of hydrocarbons of various molecular weights, and other organic compounds, that are found in geologic formations beneath the Earth's surface. Because most hydrocarbons are lighter than rock or water, they often migrate upward through adjacent rock layers until either reaching the surface or becoming trapped within porous rocks (known as reservoirs) by impermeable rocks above. However, the process is influenced by underground water flows, causing oil to migrate hundreds of kilometres horizontally or even short distances downward before becoming trapped in a reservoir.

When hydrocarbons are concentrated in a trap, an oil field forms, from which the liquid can be extracted by drilling and pumping.

The down hole pressures experienced at the rock structures change depending upon the depth and the characteristic of the source rock.

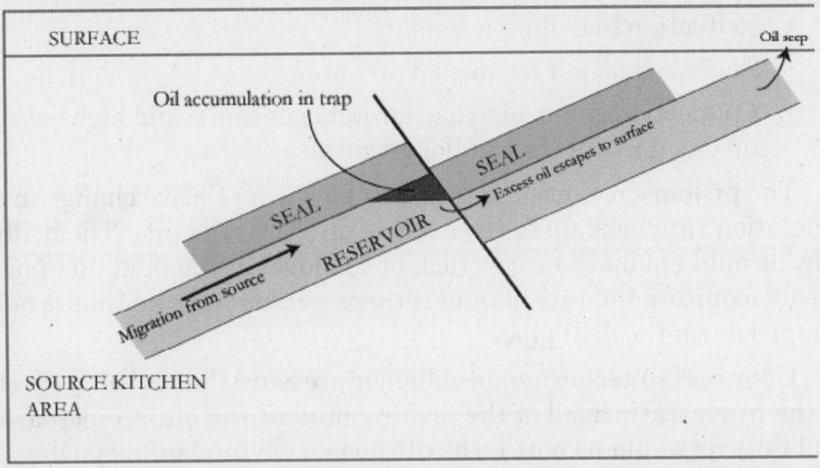

Figure: *A petroleum trap. An irregularity (the* trap*) in a layer of impermeable rocks (the* seal*) retains upward-flowing petroleum, forming a reservoir.*

Formation Kick

The downhole fluid pressures are controlled in modern wells through the balancing of the hydrostatic pressure provided by the mud used. Should the balance of the drilling mud pressure be incorrect then formation fluids (oil, natural gas and/or water) begin to flow into the wellbore and up the annulus (the space between the outside of the drill string and the walls of the open hole or the inside of the last casing string set), and/or inside the drill pipe.

This is commonly called a *kick*. If the well is not shut in (common term for the closing of the blow-out preventer valves), a kick can quickly escalate into a blowout when the formation fluids reach the surface, especially when the influx contains gas that expands rapidly as it flows up the wellbore, further decreasing the effective weight of the fluid. In other petroleum engineering words, the formation pore pressure gradient exceeds the mud pressure gradient, even in some cases when the Equivalent Circulating Density ECD is imposed with the mud pumps on the rig.

Additional mechanical barriers such as blowout preventers (BOPs) can be closed to isolate the well while the hydrostatic balance is regained through circulation of fluids in the well.

Early warning signs of a well kick are:

- Sudden change in drilling rate;
- Change in surface fluid rate;
- Change in pump pressure;
- Reduction in drillpipe weight;
- Surface mud cut by gas, oil or water;
- Connection gases, high background gas units, and high bottoms up gas units in the mudlogging unit.

The primary means of detecting a kick is a relative change in the circulation rate back up to the surface into the mud pits. The drilling crew or mud engineer keeps track of the level in the mud pits and/or closely monitors the rate of mud returns versus the rate that is being pumped down the drill pipe.

Upon encountering a zone of higher pressure than is being exerted by the hydrostatic head of the drilling mud at the bit, an increase in mud returns would be noticed as the formation fluid influx pushes the drilling mud toward the surface at a higher rate. Conversely, if the rate of returns is slower than expected, it means that a certain amount of the mud is being lost to a thief zone somewhere below the last casing shoe. This does not necessarily result in a kick (and may never become one); however, a drop in the mud level might allow influx of formation fluids from other zones if the hydrostatic head at depth is reduced to less than that of a full column of mud.

Well Control

The first response to detecting a kick would be to isolate the wellbore from the surface by activating the blow-out preventers and closing in the well. Then the drilling crew would attempt to circulate in a heavier *kill fluid* to increase the hydrostatic pressure (sometimes with the assistance of a well control company). In the process, the influx fluids will be slowly circulated out in a controlled manner, taking care not to allow any gas to accelerate up the wellbore too quickly by controlling casing pressure with chokes on a predetermined schedule.

This effect will be minor if the influx fluid is mainly salt water. And with an oil-based drilling fluid it can be masked in the early stages of controlling a kick because gas influx may dissolve into the oil under pressure at depth, only to come out of solution and expand rather rapidly as the influx nears the surface. Once all the contaminant has been circulated out, the casing pressure should have reached zero.

Types of Blowouts

Well blowouts can occur during the drilling phase, during well testing, during well completion, during production, or during workover activities.

Surface Blowouts

Blowouts can eject the drill string out of the well, and the force of the escaping fluid can be strong enough to damage the drilling rig. In addition to oil, the output of a well blowout might include sand, mud, rocks, drilling fluid, natural gas, water, and other substances.

Blowouts will often be ignited by an ignition source, from sparks from rocks being ejected, or simply from heat generated by friction. A well control company will then need to extinguish the well fire or cap the well, and replace the casing head and hangars. The flowing gas may contain poisonous hydrogen sulfide and the oil operator might decide to ignite the stream to convert this to less hazardous substances. Sometimes, blowouts can be so forceful that they cannot be directly brought under control from the surface, particularly if there is so much energy in the flowing zone that it does not deplete significantly over the course of a blowout. In such cases, other wells (called relief wells) may be drilled to intersect the well or pocket, in order to allow kill-weight fluids to be introduced at depth. When first drilled in the 1930s relief well were drilled to inject water into the main drill well hole. Contrary to what might be inferred from the term, such wells generally are not used to help relieve pressure using multiple outlets from the blowout zone.

Subsea Blowouts

Subsea wells have the wellhead and pressure control equipment located on the seabed. They vary from depths of 10 feet (3.0 m) to 8,000 feet (2,400 m). It is very difficult to deal with a blowout in very deep water because of the remoteness and limited experience with this type of situation. The Deepwater Horizon well blowout in the Gulf of Mexico in April 2010, in 5,000 feet (1,500 m) water depth, is the deepest subsea well blowout to date.

Underground Blowouts

An underground blowout is a special situation where fluids from high pressure zones flow uncontrolled to lower pressure zones within the wellbore. Usually this is from deeper higher pressure zones to shallower lower pressure formations. There may be no escaping fluid flow at the wellhead. Underground blowouts can be very difficult to bring under control, and if left unchecked the fluids may find their way to the surface or ocean floor nearby.

Blowout Control Expertise

Myron M. Kinley was a pioneer in fighting oil well fires and blowouts. He developed many patents and designs for the tools and techniques of oil firefighting. His father, Karl T. Kinley, attempted to extinguish an oil well fire with the help of a massive explosion — a method that remains a common technique for fighting oil fires. The first oil well put out with explosives by Myron Kinley and his father, was in 1913. Kinley would later form the M.M. Kinley Company in 1923.

Paul N. "Red" Adair joined the M.M. Kinley Company in 1946, and worked 14 years with Myron Kinley before starting his own company, Red Adair Co., Inc., in 1959. Asger "Boots" Hansen and Edward Owen "Coots" Matthews also begin their careers under Kinley.

Red Adair co. has helped in controlling many offshore blowouts, including;

- CATCO fire in the Gulf of Mexico in 1959
- "The Devil's Cigarette Lighter" in 1962 in Gassi Touil, Algeria, in the Sahara Desert
- The Ixtoc I oil spill in Mexico's Bay of Campeche in 1979
- The Piper Alpha disaster in the North Sea in 1988
- The Kuwaiti oil fires following the Gulf War in 1991.

In 1994, Adair retired and sold his company to Global Industries. Management of Adair's company left and created International Well Control (IWC). In 1997, they would buy the company Boots & Coots International Well Control, Inc., which was founded by two former lieutenants of Red Adair in 1978.

Methods of Quenching Blowouts

Although several experimental methods exist which attempt to capture as much oil as possible from a blown out well, they are very far from perfect, capturing between 20% - 50% of the leaking oil, by optimistic estimates. Ideally, the well could be made to stop gushing oil entirely - thus putting a stop to the cumulating pollution.

Use of Nuclear Explosions

On Sep. 30, 1966 the Soviet Union in Urt-Bulak, an area about 80 kilometres from Bukhara, Uzbekistan, experienced blowouts on five natural gas wells. It was claimed in Komsomoloskaya Pravda that after years of burning uncontrollably they were able to stop them entirely. The Soviets lowered a specially made 30 kiloton nuclear bomb into a 6 kilometres (20,000 ft) borehole drilled 25 to 50 metres (82 to 160 ft) away from the original (rapidly leaking) well. A nuclear explosive was deemed necessary because conventional explosive both lacked the necessary power and would also require a great deal more space underground.

When the bomb was set off, it proceeded to crush the original pipe that was carrying the gas from the deep reservoir to the surface, as well as to glassify all the surrounding rock. This caused the leak and fire at the surface to cease within approximately one minute of the explosion, and proved over the years to have been a permanent solution. A second attempt on a similar well was not as successful and other tests were for such experiments as oil extraction enhancement (Stavropol, 1969) and the creation of gas storage reservoirs (Orenburg, 1970). As nuclear historian Robert S. Norris notes in the Times, all these Soviet nuclear blasts were on land and did not involve oil. That being the case, the general principles involved are the same.

Differential Sticking

Differential sticking is a problem that occurs when drilling a well with a greater well bore pressure than formation pressure, as is usually the case. The drill pipe is pressed against the wellbore wall so that part of its circumference will see only reservoir pressure, while the rest will continue to be pushed by wellbore pressure. As a result the pipe becomes stuck to the wall, and can require millions of pounds of force to remove, which may prove impossible.

In many cases the drilling fluid (mud) weight is simply reduced, thus relieving the pressure difference and releasing the stuck pipe string. Should this option be unavailable, as in sour gas wells, a speciality fishing company is called to retrieve the stuck pipe or 'fish'. Many options exist once a fishing company is on site: oil or nitrogen may be pumped down the well, or the fish may be 'washed over' using a carbide shoe on a string of washpipe. Jarring is not usually attempted with differential sticking due to the massive amount of pressure that holds the pipe in place.

Directional Drilling

Directional drilling (or slant drilling) is the practice of drilling non-vertical wells. It can be broken down into three main groups:

Oilfield Directional Drilling, Utility Installation Directional Drilling (or H.D.D. or Horizontal Directional Drilling) Directional boring, and in-seam directional drilling to extract coal bed methane.

History

Many prerequisites enabled this suite of technologies to become productive. Probably, the first requirement was the realisation that oil wells, or water wells, are not necessarily vertical. This realisation was quite slow, and did not really grasp the attention of the oil industry until the late 1920s when there were several lawsuits alleging that wells drilled from a rig on one property had crossed the boundary and were penetrating a reservoir on an adjacent property. Initially, proxy evidence such as production changes in other wells was accepted, but such cases fueled the development of small diametre tools capable of surveying wells during drilling.

Measuring the inclination of a wellbore (its deviation from the vertical) is comparatively simple, requiring only a pendulum. Measuring the azimuth (direction with respect to the geographic grid in which the wellbore is running from the vertical), however, was more difficult. In certain circumstances, magnetic fields could be used, but could be influenced by metalwork used inside wellbores, as well as the metalwork used in drilling equipment.

The next advance was in the modification of small gyroscopic compasses by the Sperry Corporation, which was making similar compasses for aeronautical navigation. Sperry did this under contract to Sun Oil (which was involved in a lawsuit as described above), and a spin-off company "Sperry Sun" was formed, which brand continues to this day, absorbed into Halliburton. Three components are measured at any given point in a wellbore in order to determine its position: the depth of the point along the course of the borehole (measured depth), the inclination at the point, and the magnetic azimuth at the point. These three components combined are referred to as a "survey". A series of consecutive surveys are needed to track the progress and location of a wellbore. Many of the earliest innovations such as photographic single shot technology and crow's feet baffle plates for landing survey tools were developed by Robert Richardson, an independent directional driller who first drilled in the 1940s and was still working in 2008.

Prior experience with rotary drilling had established several principles for the configuration of drilling equipment down hole ("Bottom Hole Assembly" or "BHA") that would be prone to "drilling crooked hole" (i.e., initial accidental deviations from the vertical would be increased). Counter-experience had also given early directional drillers ("DD's") principles of BHA design and drilling practice that would help bring a crooked hole nearer the vertical.

In 1934, H. John Eastman of Long Beach, California, became a pioneer in directional drilling when he and George Failing of Enid, Oklahoma, saved the Conroe, Texas, oil field. Failing had recently patented a portable drilling truck. He had started his company in 1931 when he mated a drilling rig to a truck and a power take-off assembly. The innovation allowed rapid drilling of a series of slanted wells.

This capacity to quickly drill multiple relief wells and relieve the enormous gas pressure was critical to extinguishing the Conroe fire. (E&P, "Making a hole was hard work," Kris Wells, American Oil & Gas Historical Society Contributing Editor, Nov. 1, 2006 and "Technology and the Conroe Crater"). In a May, 1934, *Popular Science Monthly* article, it was stated that "Only a handfull of men in the world have the strange power to make a bit rotating a mile below ground at the end of a steel drill pipe, snake its way in a curve or around a dog leg angle, to reach a desired object." Eastman Whipstock, Inc., would become the world's largest directional company in 1973.

Combined, these survey tools and BHA designs made directional drilling possible, but it was perceived as arcane. The next major advance was in the 1970s, when downhole drilling motors (aka mud motors, driven by the hydraulic power of drilling mud circulated down the drill string) became common. These allowed the bit to be rotated on the bottom of the hole, while most of the drill pipe was held stationary. A piece of bent pipe (a "bent sub") between the stationary drill pipe and the top of the motor allowed the direction of the wellbore to be changed without needing to pull all the drill pipe out and place another whipstock.

Coupled with the development of Measurement While Drilling tools (using mud pulse telemetry or EM telemetry, which allows tools down hole to send directional data back to the surface without disturbing drilling operations), directional drilling became easier. Certain profiles could not be drilled without the drill string rotating at all times.

Benefits

Wells are drilled directionally for several purposes:

- Increasing the exposed section length through the reservoir by drilling through the reservoir at an angle
- Drilling into the reservoir where vertical access is difficult or not possible. For instance an oilfield under a town, under a lake, or underneath a difficult to drill formation

- Allowing more wellheads to be grouped together on one surface location can allow fewer rig moves, less surface area disturbance, and make it easier and cheaper to complete and produce the wells. For instance, on an oil platform or jacket offshore, 40 or more wells can be grouped together. The wells will fan out from the platform into the reservoir(s) below. This concept is being applied to land wells, allowing multiple subsurface locations to be reached from one pad, reducing costs.
- Drilling along the underside of a reservoir-constraining fault allows multiple productive sands to be completed at the highest stratigraphic points.
- Drilling a "relief well" to relieve the pressure of a well producing without restraint (a "blow out"). In this scenario, another well could be drilled starting at a safe distance away from the blow out, but intersecting the troubled wellbore. Then, heavy fluid (kill fluid) is pumped into the relief wellbore to suppress the high pressure in the original wellbore causing the blowout.

Most directional drillers are given a well path to follow that is predetermined by engineers and geologists before the drilling commences. When the directional driller starts the drilling process, periodic surveys are taken with a downhole instrument to provide survey data (inclination and azimuth) of the well bore. These pictures are typically taken at intervals between 30–500 feet, with 90 feet common during active changes of angle or direction, and distances of 200–300 feet being typical while "drilling ahead" (not making active changes to angle and direction). During critical angle and direction changes, especially while using a downhole motor, an MWD (Measurement while drilling) tool will be added to the drill string to provide continuously updated measurements that may be used for (near) real-time adjustments.

These data indicate if the well is following the planned path and whether the orientation of the drilling assembly is causing the well to deviate as planned. Corrections are regularly made by techniques as simple as adjusting rotation speed or the drill string weight (weight on bottom) and stiffness, as well as more complicated and time consuming methods, such as introducing a downhole motor. Such pictures, or surveys, are plotted and maintained as an engineering and legal record describing the path of the well bore. The survey pictures taken while drilling are typically confirmed by a later survey in full of the borehole, typically using a "multi-shot camera" device.

The multi-shot camera advances the film at time intervals so that by dropping the camera instrument in a sealed tubular housing inside the drilling string (down to just above the drilling bit) and then withdrawing the drill string at time intervals, the well may be fully surveyed at regular depth intervals (approximately every 90 feet being common, the typical length of 2 or 3 joints of drill pipe, known as a stand, since most drilling rigs "stand back" the pipe withdrawn from the hole at such increments, known as "stands").

Drilling to targets far laterally from the surface location requires careful planning and design. The current record holders manage wells over 10 km (6.2 mi) away from the surface location at a true vertical depth (TVD) of only 1,600–2,600 m (5,200–8,500 ft).

Disadvantages

Until the arrival of modern downhole motors and better tools to measure inclination and azimuth of the hole, directional drilling and horizontal drilling was much slower than vertical drilling due to the need to stop regularly and take time consuming surveys, and due to slower progress in drilling itself (lower rate of penetration). These disadvantages have shrunk over time as downhole motors became more efficient and semi-continuous surveying became possible. What remains is a difference in operating costs: for wells with an inclination of less than 40 degrees, tools to carry out adjustments or repair work can be lowered by gravity on cable into the hole. For higher inclinations, more expensive equipment has to be mobilised to push tools down the hole. Another disadvantage of wells with a high inclination was that prevention of sand influx into the well was less reliable and needed higher effort. Again, this disadvantage has diminished such that, provided sand control is adequately planned, it is possible to carry it out reliably.

Stealing Oil

In 1990, Iraq accused Kuwait of stealing Iraq's oil through slant drilling. The United Nations redrew the border after the 1991 Gulf war that liberated Kuwait from a seven-month Iraqi occupation under former leader Saddam Hussein. It placed 11 oil wells, some farms and an old naval base that used to be in Iraq on the Kuwaiti side.

In the mid-twentieth century, a slant-drilling scandal occurred in the huge East Texas Oil Field.

Measurement While Drilling

MWD stands for Measurement While Drilling in the oil & gas industry. The simplest way to describe MWD is to relate it to the measurements a pilot takes. A pilot needs to know the direction they are flying (North, South, East, or West), the angle they are fly at (up, down, or horizontal), and what type of skies they will be flying through (rough, choppy, cloudy, rainy, etc). Like a pilot a directional driller needs to know these items about the ground formations that they are drilling through. MWD provides this information. Previous to MWD measurements were taken at various parts of the drilling process, but MWD has allowed these measurements to be sent to the surface continuously while the hole is being drilled. This allows for faster drilling, more accurate drilling, and safer drilling.

Beyond the basic concept MWD is a system developed to perform drilling related measurements downhole that are transmit to the surface while drilling a well. MWD tools are installed as part of the bottom hole assembly (BHA) near the drill bit. The tools are either contained inside a thick walled, drill collar (drill collars are typically used to add weight for drilling) or they are built directly into the collars at a factory prior to arriving on the drilling location.

MWD systems can take several measurements such as Gamma Ray, compass direction (shown as azimuth), tool face (the direction that your bit is pointing), borehole pressure, temperature, vibration, shock, torque, etc. Some advanced MWD tools can even measure formation pressure and take formation samples. The MWD also provides the means of communication for operating rotary steering tools (RSTs).

The measured results are stored in MWD tools and some of the results can be transmitted digitally to surface using mud pulser telemetry through the mud or other advanced technology such as electromagnetic(EM) frequency communications or wired drill pipe.

Types of Information Transmitted

Directional Information: MWD tools are generally capable of taking directional surveys in real time. The tool uses accelerometres and magnetometres to measure the inclination and azimuth of the wellbore at that location, and they then transmit that information to the surface. With a series of surveys at appropriate intervals (anywhere from every 30 ft (ie 10m) to every 500 ft), the location of the wellbore can be calculated.

By itself, this information allows operators to prove that their well does not cross into areas that they are not authorized to drill. However, due to the cost of MWD systems, they are not generally used on wells intended to be vertical. Instead, the wells are surveyed after drilling through the use of Multishot Surveying Tools lowered into the drillstring on slickline or wireline.

The primary use of real-time surveys is in Directional Drilling. For the Directional Driller to steer the well towards a target zone, he must know where the well is going, and what the effects of his steering efforts are.

MWD tools also generally provide toolface measurements to aid in directional drilling using downhole mud motors with bent subs or bent housings.

Drilling Mechanics Information

MWD tools can also provide information about the conditions at the drill bit. This may include:

- Rotational speed of the drillstring
- Smoothness of that rotation
- Type and severity of any vibration downhole
- Downhole temperature
- Torque and Weight on Bit, measured near the drill bit
- Mud flow volume.

Use of these information can allow the operator to drill the well more efficiently, and to ensure that the MWD tool and any other downhole tools, such as Mud Motors, Rotary Steerable Systems, and LWD tools, are operated within their technical specifications to prevent tool failure. This information also is valuable to Geologists responsible for the well information about the formation which is being drilled.

Formation Properties

Many MWD tools, either on their own, or in conjunction with separate Logging While Drilling tools, can take measurements of formation properties. At the surface, these measurements are assembled into a log, similar to one obtained by wireline logging.

LWD Logging While Drilling tools are able to measure a suite of geological characteristics including- density, porosity, resistivity, acoustic-caliper, inclination at the drill bit (NBI), magnetic resonance and formation pressure.

The MWD tool allows these measurements to be taken and evaluated while the well is being drilled. This makes it possible to perform Geosteering, or Directional Drilling based on measured formation properties, rather than simply drilling into a preset target.

Most MWD tools contain an internal Gamma Ray sensor to measure natural Gamma Ray values. This is because these sensors are compact, inexpensive, reliable, and can take measurements through unmodified drill collars. Other measurements often require separate Logging While Drilling tools, which communicate with the MWD tools downhole through internal wires. Measurement while drilling can be cost-effective in exploration wells, particularly in areas of the Gulf of Mexico where wells are drilled in areas of salt diapirs. The resistivity log will detect penetration into salt, and early detection prevents salt damage to bentonite drilling mud.

Data Transmission Methods

Mud Pulse Telemetry

This is the most common method of data transmission used by MWD (Measurement While Drilling) tools. Downhole a valve is operated to restrict the flow of the drilling mud (slurry) according to the digital information to be transmitted. This creates pressure fluctuations representing the information. The pressure fluctuations propagate within the drilling fluid towards the surface where they are received from pressure sensors. On the surface, the received pressure signals are processed by computers to reconstruct the information. The technology is available in three varieties - *positive* pulse, *negative* pulse, and *continuous wave.*

Positive Pulse: Positive Pulse tools briefly close and open the valve to restrict the mud flow within the drill pipe. This produces an increase in pressure that can be seen at surface. Line codes are used to represent the digital information in form of pulses.

Negative Pulse: Negative pulse tools briefly open and close the valve to release mud from inside the drillpipe out to the annulus. This produces a decrease in pressure that can be seen at surface. Line codes are used to represent the digital information in form of pulses.

Continuous Wave: Continuous wave tools gradually close and open the valve to generate sinusoidal pressure fluctuations within the drilling fluid. Any digital modulation scheme with a continuous phase can be used to impose the information on a carrier signal. The most widely used modulation scheme is continuous phase modulation.

When underbalanced drilling is used, mud pulse telemetry can become unusable. This is because usually in order to reduce the equivalent density of the drilling mud a compressible gas is injected into the mud. This causes high signal attenuation which drastically reduces the ability of the mud to transmit pulsed data. In this case it is necessary to use methods different from mud pulse telemetry, such as electromagnetic waves propagating through the formation or wired drill pipe telemetry.

Current mud pulse telemetry technology offers a bandwidths of up to 40 bps. The data rate drops with increasing length of the wellbore and is typically as low as 1.5 bps - 3.0 bps. (bits per second) at a depth of 35,000 ft - 40,000 ft (10668 m - 12192 m).

Surface to down hole communication is typically done via changes to drilling parameters, i.e. change of the rotation speed of the drill string or change of the mud flow rate. Making changes to the drilling parameters in order to send information can require interruption of the drilling process, which is unfavourable due to the fact that it causes non-productive time.

Electromagnetic Telemetry (EM Tool)

These tools incorporate an electrical insulator in the drillstring. To transmit data the tool generates an altered voltage difference between the top part (the main drillstring, above the insulator), and the bottom part (the drill bit, and other tools located below the insulator of the MWD tool). On surface a wire is attached to the wellhead, which makes contact with the drillpipe at the surface. A second wire is attached to a rod driven into the ground some distance away. The wellhead and the ground rod form the two electrodes of a dipole antenna. The voltage difference between the two electrodes is the receive signal that is decoded by a computer.

The EM tool generates voltage differences between the drillstring sections in the pattern of very low frequency (2–12 Hz) waves. The data is imposed on the waves through digital modulation.

This system generally offers data rates of up to 10 bits per second. In addition, many of these tools are also capable of receiving data from the surface in the same way, while mud pulse-based tools rely on changes in the drilling parameters, such as rotation speed of the drillstring or the mud flow rate, to send information from the surface to downhole tools. Making changes to the drilling parameters in order to send information to the tools generally interrupts the drilling process, causing lost time.

Compared to mud pulse telemetry, electronic pulse telemetry is more effective in certain specialised situations, such as underbalanced drilling or when using air as drilling fluid. However, it generally falls short when drilling exceptionally deep wells, and the signal can lose strength rapidly in certain types of formations, becoming undetectable at only a few thousand feet of depth.

Wired Drill Pipe

Several oilfield service companies are currently developing wired drill pipe systems. These systems use electrical wires built into every component of the drillstring, which carry electrical signals directly to the surface. These systems promise data transmission rates orders of magnitude greater than anything possible with mud pulse or electromagnetic telemetry, both from the downhole tool to the surface, and from the surface to the downhole tool. The IntelliServ wired pipe network, offering data rates upwards of 1 megabit per second, became commercial in 2006. Representatives from BP America, StatoilHydro, Baker Hughes INTEQ, and Schlumberger presented three success stories using this system, both onshore and offshore, at the March, 2008 SPE/IADC Drilling Conference in Orlando, Florida.

Retrievable Tools

MWD tools may be semi-permanently mounted in a drill collar (only removable at servicing facilities), or they may be self-contained and wireline retrievable. Retrievable tools, sometimes known as *Slim Tools*, can be retrieved and replaced using wireline through the drill string. This generally allows the tool to be replaced much faster in case of failure, and it allows the tool to be recovered if the drillstring becomes stuck. Retrievable tools must be much smaller, usually about 2 inches or less in diametre, though their length may be 20 feet or more. The small size is necessary for the tool to fit through the drillstring, however, it also limits the tool's capabilities. For example, slim tools are not capable of sending data at the same rates as collar mounted tools, and they are also more limited in their ability to communicate with and supply electrical power to other LWD tools.

Collar-mounted tools, also known as *Fat Tools*, cannot generally be removed from their drill collar at the wellsite. If the tool fails, the entire drillstring must be pulled out of the hole to replace it. However, without the need to fit through the drillstring, the tool can be larger and more capable.

The ability to retrieve the tool via wireline is often useful. For example, if the drillstring becomes stuck in the hole, then retrieving the tool via wireline will save a substantial amount of money compared to leaving it in the hole with the stuck portion of the drillstring. However, there are some limitations on the process.

Limitations

Retrieving a tool using wireline is not necessarily faster than pulling the tool out of the hole. For example, if the tool fails at 1,500 ft (460 m) while drilling with a triple rig (able to trip 3 joints of pipe, or about 90 ft (30 m) feet, at a time), then it would generally be faster to pull the tool out of the hole than it would be to rig up wireline and retrieve the tool, especially if the wireline unit must be transported to the rig.

Wireline retrievals also introduce additional risk. If the tool becomes detached from the wireline, then it will fall back down the drillstring. This will generally cause severe damage to the tool and the drillstring components in which it seats, and will require the drillstring to be pulled out of the hole to replace the failed components, thus resulting in a greater total cost than pulling out of the hole in the first place. The wireline gear might also fail to latch onto the tool, or in the case of a severe failure, might bring only a portion of the tool to the surface. This would require the drillstring to be pulled out of the hole to replace the failed components, thus making the wireline operation a waste of time.

Geosteering

In the process of drilling a borehole, geosteering is the act of adjusting the borehole position (inclination and azimuth angles) *on the fly* to reach one or more geological targets. These changes are based on geological information gathered while drilling.

Description

From 2D and 3D models of underground substructures, deviated wells (2D and 3D) are planned in advance to achieve specific goals: exploration, fluids production, fluids injection or technical.

A *well plan* is a continuous succession of straight and curved lines representing the geometrical figure of the expected well path. A well plan is always projected on vertical and horizontal maps.

While the borehole is being drilled according to the well plan, new geological information is gathered from mud logging, Measurement While Drilling and Logging While Drilling. These usually show some differences from what is expected from the model. As the model is continuously updated with the new geological information (formation evaluation) and the borehole position (*well deviation survey*), changes start to appear in the geological substructures and can lead to the well plan being updated to reach the corrected geological targets.

Drilling Engineering

Drilling engineers design and implement procedures to drill wells as safely and economically as possible. They work closely with the drilling contractor, service contractors, and compliance personnel, as well as with geologists and other technical specialists. The drilling engineer has the responsibility for ensuring that costs are minimised while getting information to evaluate the formations penetrated, protecting the health and safety of workers and other personnel, and protecting the environment.

Overview

The planning phases involved in drilling an oil or gas well typically involve estimating the value of sought reserves, estimating the costs to access reserves, acquiring property by a mineral lease, a geological survey, a well bore plan, and a layout of the type of equipment required to reach the depth of the well. Drilling engineers are in charge of the process of planning and drilling the wells. Their responsibilities include:

1. Designing well programs (e.g., casing sizes and setting depths) to prevent blowouts (uncontrolled well-fluid release) while allowing adequate formation evaluation.
2. Designing or contributing to the design of casing strings and cementing plans, directional drilling plans, drilling fluids programs, and drill string and drill bit programs.
3. Specifying equipment, material and ratings and grades to be used in the drilling process.
4. Providing technical support and audit during the drilling process.
5. Performing cost estimates and analysis.
6. Developing contracts with vendors.

Drilling engineers are often degreed as petroleum engineers, although they may come from other technical disciplines (e.g., mechanical engineering or geology) and subsequently be trained by an oil and gas company. They also may have practical experience as a rig hand or mudlogger or mud engineer.

Drilling Fluid Invasion

Drilling fluid invasion is a process that occurs in a well being drilled with higher wellbore pressure than formation pressure. The liquid component of the drilling fluid (known as the mud filtrate) continues to "invade" the porous and permeable formation until the solids present in the mud, commonly bentonite, clog enough pores to form a mud cake capable of preventing further invasion.

If invasion is severe enough, and reservoir pressures are unable to force the fluid and associated particles out entirely when the well starts producing, the amount of oil and gas a well can produce can be permanently reduced. This is especially true when a process called phase trapping occurs. This is when a fluid enters a formation that is below its irreducible saturation of that fluid. Once the fluid is present, it is held in place by capillary forces and usually can not be removed.

Invasion also has significant implications for well logging. In many cases the "depth of investigation" of a well logging tool is only a few inches (or even less for things like sonic logs), and it is quite possible that drilling fluid has invaded beyond this depth. Therefore the readings that are taken that are influenced by formation fluids are measuring mud filtrate properties rather than formation (in situ) fluid properties.

Drill Stem Test

A drill stem test (DST) is a procedure for isolating and testing the surrounding geological formation through the drill stem. The test is a measurement of pressure behaviour at the drill stem and is a valuable way to obtain important sampling information on the formation fluid and to establish the probability of commercial production.

In oil and natural gas extraction, the drill stem includes the drill pipe, drill collars, bottomhole assembly, and drill bit. During normal drilling, fluid is pumped through the drill stem and out the drill bit. In a drill stem test, the drill bit is removed, a drill stem test tool is added, and fluid from the formation is recovered through the drill stem, while several measurements of pressure are being made.

The basic drill stem test tool consists of a packer or packers, valves or ports that may be opened and closed from the surface, and two or more pressure-recording devices. (A packer is an expanding plug which can be used to seal off sections of the open or cased well, to isolate them for testing.) The drill stem test tool is lowered on the drill pipe to the zone to be tested. The packer or packers are set to isolate the zone from the drilling fluid column, the tester valve is opened, and testing begins.

History

Working in El Dorado, Arkansas, in the 1920s, E.C. Johnston and his brother M.O. Johnston developed the first drill stem tester and ran the first commercial drill stem test in 1926. In April 1929, the Johnston

Formation Testing Corporation was granted a patent (U.S. Patent 1,709,940) and they subsequently refined the testing system in the early 1930s.

In the 1950s, Schlumberger introduced a method for testing formations using wireline. The Schlumberger formation-testing tool, placed in operation in 1953, fired a shaped charge through a rubber pad that had been expanded in the hole until it was securely fixed in the hole at the depth required. Formation fluids flowed through the perforation and connecting tubing into a container housed inside the tool. When filled, the container was closed, sealing the fluid sample at the formation pressure. The tool was then brought to the surface, where the sample could be examined. In 1956, Schlumberger acquired Johnston Testers and continues to perform drill stem tests and wireline formation tests in both open and cased holes.

Lost Circulation

Lost circulation can be one of the more serious problems that can arise during the drilling of an oil well or gas well. Circulation is said to be lost when the drilling fluid, known commonly as "mud", flows into one or more geological formations instead of returning up the annulus.

Consequences

The consequences of lost circulation can be as little as the loss of a few dollars of drilling fluid, or as disastrous as a blowout and loss of life, so close monitoring of tanks, pits, and flow from the well, to quickly assess and control lost circulation, is taught and practiced. If the fluid in the wellbore drops due to lost circulation (or any other reason), hydrostatic pressure is reduced, thus allowing a gas or fluid, which is under a higher pressure than the reduced hydrostatic pressure, to influx into the wellbore.

Another consequence of lost circulation is called "dry drilling". Dry drilling occurs when fluid is completely lost from the well bore without actual drilling coming to a stop. The effects of dry drilling can be as minor as destroying a bit to as serious as major damage to the wellbore requiring a new well to be drilled. Dry drilling can also cause severed damage to the drill string, including snapping the pipe, and the drilling rig itself.

Categories

Lost circulation falls into two main categories:

- "Minor losses" - This is where losses are between 6 and 470 barrels (1 to 75 m^3), and remains within those amounts, or is ceased, within 48 hours.

- "Severe losses" - This is where losses are greater than 470 barrels (75 m^3), or it takes greater than 48 hours to control or cease the lost circulation.

Total losses may also occur, where the return of fluids is completely lost to the surface. This may fall into either the minor or severe losses categories, depending on the amount of losses and the time involved in regaining circulation.

Control

Although preferred, ceasing lost circulation completely is not always possible or required. Controlled losses allows drilling to continue while keeping the wellbore full, preventing an influx of gas or fluid into the wellbore, known as a" kick", which can lead to a blowout.

A number of options are available when lost circulation occurs, depending on the severity.

- Minor losses may be controlled by increasing the viscosity of the fluid with bentonite and/ or polymers, or with the addition of other additives which typically includes sawdust.
- Severe losses will require increasing the viscosity of the fluid with bentonite and/ or polymers and the addition of other additives which typically includes sawdust.

Total losses can be regained through conventional use of increased viscosity and additives, or through use of unconventional methods such as pumping of golf balls, tree branches, rags, additive sacks, and other items in conjunction with, or followed by, a high viscosity fluid. If total losses occur and circulation cannot be regained, several options are available, depending on the operational requirements and depth being drilled in relation to desired production geological zones. Continuing drilling while pumping drilling fluid is one option, though continued drilling while pumping water is less costly and more often used. Sometimes the cuttings from continued drilling will aid in reducing or stop losses altogether. A third option is to cement the zone where the losses occurred, and to drill through the cement and continue drilling the well. This third option is very often the most cost effective if severe losses occur, as lost circulation can sometimes not be controlled with conventional or unconventional methods.

Additives

- The most common additive used to control or cease lost circulation is bentonite, in that it inherently will seal small holes or fractures. Bentonite, in higher concentrations, increases viscosity, and therefore slows the fluid flow into the surrounding rock. Although bentonite is the most common additive used, it sometimes is not used at all, depending on the current fluid being used and the current depth of drilling in relation to desired production zones.

- Polymers are also sometimes used to increase the viscosity. Though these are more costly, they are more compatible with several types of fluid systems.
- When it comes to additives which physically plug or seal the losses, there are a great deal of them. Sawdust, flaked cellophane, crushed or ground gypsum are all used.
- Other common, and cheaper additives, that are used are shredded newspaper and cotton seed hulls. Cotton seed hulls are less preferred as they may cause wear to pump swabs and springs. Both of these are generally only used when either fresh or brine water is being used for the drilling fluid.
- When drilling in salt formations, brine water is typically used as it is more difficult to dissolve more of the salt preventing the formation of "washouts". Washouts not only contribute to loss of circulation, but can jeopardize the integrity of the wellbore itself.

Additive Considerations

Several factors are considered in what additives are used:

- Hole size currently being drilled.
- Drilling fluid in use. The additives must be compatible.
- Depth of the well in regards to geological stability.
- Depth of well in relation to the desired production zones. Plugging a production zone is not a desired outcome.
- Drill bit nozzles sizes. If the additive(s) will not go through the drill bit, they cannot be used.
- Other drill string mechanical equipment such as a mud motor or MWD tools. If the additive(s) will not go through the drill string, they cannot be used.

Tracer Use in the Oil Industry

Tracers are used in the oil industry in order to qualitatively or quantitatively gauge how fluid flows through the reservoir, as well as being a useful tool for estimating residual oil saturation. Tracers can be used in either interwell tests or single well tests. In interwell tests, the tracer is injected at one well along with the carrier fluid (water in a waterflood or gas in a gasflood) and detected at a producing well after some period of time, which can be anything from days to years. In single well tests, tracer is injected into the formation from a well and then produced out the same well. The delay between a tracer that does not react with the formation (a conservative tracer) and one that does (a partitioning tracer) will give an indication of residual oil saturation, a piece of information that is difficult to acquire by other means.

Tracers can be radioactive or chemical, gas or liquid and have been used extensively in the oil industry and hydrology for decades.

Underbalanced Drilling

Underbalanced drilling, or UBD, is a procedure used to drill oil and gas wells where the pressure in the wellbore is kept lower than the fluid pressure in the formation being drilled. As the well is being drilled, formation fluid flows into the wellbore and up to the surface. This is the opposite of the usual situation, where the wellbore is kept at a pressure above the formation to prevent formation fluid entering the well. In such a conventional "overbalanced" well, the invasion of fluid is considered a kick, and if the well is not shut-in it can lead to a blowout, a dangerous situation. In underbalanced drilling, however, there is a "rotating head" at the surface - essentially a seal that diverts produced fluids to a separator while allowing the drill string to continue rotating.

If the formation pressure is relatively high, using a lower density mud will reduce the well bore pressure below the pore pressure of the formation. Sometimes an inert gas is injected into the drilling mud to reduce its equivalent density and hence its hydrostatic pressure throughout the well depth. This gas is commonly nitrogen, as it is non-combustible and readily available, but air, reduced oxygen air, processed flue gas and natural gas have all been used in this fashion.

Kinds of Underbalanced Drilling

There are several kinds of underbalanced drilling. The most common are listed below.

- *Dry Air:* This is also known as dusting. Here air compressors combined with a booster (which takes the head from the compressors and increases the pressure of the air, but does not increase the volume of air going down hole) are used and the only fluid injected into the well is a small amount of oil to reduce corrosion.
- *Mist:* A small amount of foaming agent (soap) is added into the flow of air. Fine particles of water and foam in an atmosphere of air bring cuttings back to the surface.

- *Foam:* A larger amount of foaming agent is added into the flow. Bubbles and slugs of bubbles in an atmosphere of mist bring cuttings back to the surface.
- *Stable Foam:* An even larger amount of foaming agent is added into the flow. This is the consistency of a shaving cream.
- *Airlift:* Slugs and bubbles of air in a matrix of water, soap can or can not be added into the fluid flow of air.
- *Aerated Mud:* Air or another gas is injected into the flow of drilling mud. Degassing units are required to remove air before it can be recirculated.

Advantages

Underbalanced wells have several advantages over conventional drilling including:

- Eliminated formation damage. In a conventional well, drilling mud is forced into the formation in a process called invasion, which frequently causes formation damage - a decrease in the ability of the formation to transmit oil into the wellbore at a given pressure and flow rate. It may or may not be repairable. In underbalanced drilling, if the underbalanced state is maintained until the well becomes productive, invasion does not occur and formation damage can be completely avoided.
- Increased Rate of Penetration (ROP). With less pressure at the bottom of the wellbore, it is easier for the drill bit to cut and remove rock.
- Reduction of lost circulation. Lost circulation is when drilling mud flows into the formation uncontrollably. Large amounts of mud can be lost before a proper mud cake forms, or the loss can continue indefinitely. If the well is drilled underbalanced, mud will not enter the formation and the problem can be avoided.
- Differential sticking is eliminated. Differential sticking is when the drill pipe is pressed against the wellbore wall so that part of its circumference will see only reservoir pressure, while the rest will continue to be pushed by wellbore pressure. As a result the pipe becomes stuck to the wall, and can require thousands of pounds of force to remove, which may prove impossible. Because the reservoir pressure is greater than the wellbore pressure in UBD, the pipe is pushed away from the walls, eliminating differential sticking.
- Formation damage Some rock formation have a reactive tendency to water. When drillmud is used the water in the drill

mud reacts with the formation (mostly clay) and inheriently causes a formation damage (reduction in permeability and porosity) Use of underbalanced drilling can prevent it

Disadvantages

Underbalanced drilling is usually more expensive than conventional drilling (when drilling a deviated well which requires directional drilling tools), and has safety issues of its own. Technically the well is always in a blowout condition unless a heavier fluid is displaced into the well. Air drilling requires a faster up hole volume as the cuttings will fall faster down the annulus when the compressors are taken off the hole compared to having a higher viscosity fluid in the hole. Because air is compressible mud pulse telemetry measurement while drilling (MWD) tools which require an incompressible fluid can not work. Common technologies used to eliminate this problem are either electromagnetic MWD tools or wireline MWD tools. Downhole mechanics are usually more violent also because the volume of fluid going through a downhole motor or downhole hammer is greater than an equivalent fluid when drilling balanced or over balanced because of the need of higher up hole velocities. Corrosion is also a problem, but can be largely avoided using a coating oil or rust inhibitors.

Well Logging

Well logging, also known as borehole logging is the practice of making a detailed record (a *well log*) of the geologic formations penetrated by a borehole. The log may be based either on visual inspection of samples brought to the surface (*geological* logs) or on physical measurements made by instruments lowered into the hole (*geophysical* logs). Well logging can be done during any phase of a well's history; drilling, completing, producing and abandoning. Well logging is done in boreholes drilled for the oil and gas, groundwater, minerals, geothermal, and for environmental and geotechnical studies.

Electric or Geophysical Well Logs

The oil and gas industry records rock and fluid properties to find hydrocarbon zones in the geological formations intersected by a borehole. The logging procedure consists of lowering a 'logging tool' on the end of a wireline into an oil well (or hole) to measure the rock and fluid properties of the formation. An interpretation of these measurements is then made to locate and quantify potential depth zones containing oil and gas (hydrocarbons). Logging tools developed over the years measure the electrical, acoustic, radioactive, electromagnetic, nuclear magnetic resonance, and other properties of the rocks and their contained fluids. Logging is usually performed as the logging tools are pulled out of the hole. This data is recorded either at surface (real-time mode), or downhole (memory mode) to electronic data format and then either a printed record or electronic presentation called a "well log" provided to the client. Well logging is performed at various intervals during the drilling of the well and when the total depth is drilled, which could range in depths from 150 m to 10668 m (500 ft to 35,000 ft) or more.

Electric line is the common term for the armoured, insulated cable used to conduct current to downhole tools used for well logging. Electric line can be subdivided into open hole operations and cased hole operations. Other conveyance methods for logging are logging while drilling (LWD), tractor, coiled tubing (real-time and memory), drill pipe conveyed, and slickline (memory, and with new development, some slickline telemetry capability).

Open hole operations, or reservoir evaluation, involve the deployment of tools into a freshly drilled well. As the toolstring traverses the wellbore, the individual tools gather information about the surrounding formations. A typical open hole log will have information about the density, porosity, permeability, lithology, presence of hydrocarbons, and oil and water saturation.

Cased hole operations, or production optimization, focus on optimizing the completed oil well through mechanical services and logging technologies. At this point in the well's life, the well is encased in steel pipe, cemented into the well bore and may or may not be producing. A typical cased hole log may show cement quality, production information, and formation data. Mechanical services use jet perforating guns, setting tools, and dump bailors to optimize the flow of hydrocarbons.

Wireline Tool Types

Typically the wireline tools are cylindrical in shape, usually from 1.5 to 5 inches in diametre. "Open hole" tool combinations can extend to over 100 feet long; "cased hole" tool combinations are often limited in length by the height restrictions imposed by constraints of "lubricator" pipe section required to contain the well pressure while deploying cased hole tools. There are many types of logging tools, ranging from common measurements (pressure and temperature) to advanced rock properties and fracture analysis, fluid properties in the wellbore, or formation properties extending several metres into the rock formation.

1. *With Sensors without Excitation*: There are units to measure spontaneous potential (SP), which is a voltage difference between a surface electrode and another electrode located in the downhole instrument, other instruments that measure the natural radiation from natural isotopes of potassium, thorium, etc., to measure pressure and temperature, etc.
2. *With Sources of Excitation and Sensors*: There are sensor systems consistent with a source of excitation and a sensor. In this type we find acoustic (also called sonic), electric, inductive, magnetic resonance, sensing systems, just to name a few.
3. *Instruments that Produce Some Mechanical Work, or Retrieve a Sample of Fluid or Rock to the Surface*: Devices to collect samples of rock, samples of fluid extracted from the rock, and some other mechanical devices.

Types of Electric/Electronic Logs

There are many types of electric/electronic logs and they can be categorized either by their function or by the technology that they use. "Open hole logs" are run before the oil or gas well is lined with pipe or cased. "Cased hole logs" are run after the well is lined with casing or production pipe.

Electric/electronic logs can also be divided into two general types based on what physical properties they measure. Resistivity logs measure some aspect of the specific resistance of the geologic formation. There are about 17 types of resistivity logs.

Porosity logs measure the fraction or percentage of pore volume in a volume of rock. Most porosity logs use either acoustic or nuclear technology. Acoustic logs measure characteristics of sound waves propagated through the well-bore environment. Nuclear logs utilise nuclear reactions that take place in the downhole logging instrument or in the formation. Nuclear logs include density logs and neutron logs, as well as gamma ray logs which are used for correlation. The basic principle behind the use of nuclear technology is that a neutron source placed near the formation of which the porosity is required to be measured will result in neutrons being scattered by the hydrogen atoms, largely those present in the formation fluid. Since there is little difference in the neutrons scattered by hydrocarbons or water, the porosity measured gives a figure close to the true physical porosity whereas the figure obtained from electrical resistivity measurements is that due to the conductive formation fluid. The difference between neutron porosity and electrical porosity measurements therefore indicates the presence of hydrocarbons in the formation fluid.

History

Conrad and Marcel Schlumberger, who founded Schlumberger Limited in 1926, are considered the inventors of electric well logging. Conrad developed the Schlumberger array, which was a technique for prospecting for metal ore deposits, and the brothers adopted that surface technique to subsurface applications. On September 5, 1927, a crew working for Schlumberger lowered an electric sonde or tool down a well in Pechelbronn, Alsace, France creating the first well log. In modern terms, the first log was a resistivity log that could be described as 3.5-metre upside-down lateral log.

In 1931, Henri George Doll and G. Dechatre, working for Schlumberger, discovered that the galvanometre wiggled even when no current was being passed through the logging cables down in the well. This led to the discovery of the spontaneous potential (SP) which was as important as the ability to measure resistivity. The SP effect was produced naturally by the borehole mud at the boundaries of permeable beds. By simultaneously recording SP and resistivity, loggers could distinguish between permeable oil-bearing beds and impermeable nonproducing beds. In 1940, Schlumberger invented the spontaneous potential dipmetre; this instrument allowed the calculation of the dip and direction of the dip of a layer. The basic dipmetre was later enhanced by the resistivity dipmetre (1947) and the continuous resistivity dipmetre (1952).

Oil-based mud (OBM) was first used in Rangely Field, Colorado in 1948. Normal electric logs require a conductive or water-based mud, but OBMs are nonconductive. The solution to this problem was the induction log, developed in the late 1940s.

The introduction of the transistor and integrated circuits in the 1960s made electric logs vastly more reliable. Computerization allowed much faster log processing, and dramatically expanded log data-gathering capacity. The 1970s brought more logs and computers. These included combo type logs where resistivity logs and porosity logs were recorded in one pass in the borehole. The two types of porosity logs (acoustic logs and nuclear logs) date originally from the 1940s. Sonic logs grew out of technology developed during World War II. Nuclear logging has supplemented acoustic logging, but acoustic or sonic logs are still run on some combination logging tools.

Nuclear logging was initially developed to measure the natural gamma radiation emitted by underground formations. However, the industry quickly moved to logs that actively bombard rocks with nuclear particles. The gamma ray log, measuring the natural radioactivity, was introduced by Well Surveys Inc. in 1939, and the WSI neutron log came in 1941. The gamma ray log is particularly useful as shale beds which often provide a relatively low permeability cap over hydrocarbon reservoirs usually display a higher level of gamma radiation. These logs were important because they can be used in cased wells (wells with production casing). WSI quickly became part of Lane-Wells. During World War II, the US Government gave a near wartime monopoly on open-hole logging to Schlumberger, and a monopoly on cased-hole logging to Lane-Wells. Nuclear logs continued to evolve after the war.

The nuclear magnetic resonance log was developed in 1958 by Borg Warner. Initially the NMR log was a scientific success but an engineering failure. However, the development of a continuous NMR logging tool by Numar (now a subsidiary of Halliburton) is a promising new technology.

Many modern oil and gas wells are drilled directionally. At first, loggers had to run their tools somehow attached to the drill pipe if the well was not vertical. Modern techniques now permit continuous information at the surface. This is known as logging while drilling (LWD) or measurement-while-drilling (MWD). MWD logs use mud pulse technology to transmit data from the tools on the bottom of the drillstring to the processors at the surface.

Logging While Drilling

In the 1970s, a new technique, logging while drilling (LWD), was introduced which provided similar information about the well. Instead of sensors being lowered into the well at the end of wireline cable, the sensors are integrated into the drill string and the measurements are made while the well is being drilled. While wireline well logging occurs after the drill string is removed from the well, LWD measures and transmits geological parameters while the well is being drilled. Measured data is transmitted to the surface in real time via pressure pulses in the well's mud fluid column. This mud telemetry method provides a bandwidth of less than 10 bits per second, although, as drilling through rock is a fairly slow process, data compression techniques mean that this is an ample bandwidth for real-time delivery of information. A higher sample rate of data is recorded into memory and retrieved when the drillstring is withdrawn at bit changes.

Logging Measurement Types

Logging measurements are quite sophisticated. The prime target is the measurement of various geophysical properties of the subsurface rock formations. Of particular interest are porosity, permeability, and fluid content. Porosity is the proportion of fluid-filled space found within the rock. It is this space that contains the oil and gas. Permeability is the ability of fluids to flow through the rock. The higher the porosity, the higher the possible oil and gas content of a rock reservoir. The higher the permeability, the easier for the oil and gas to flow toward the wellbore. Logging tools provide measurements that allow for the mathematical interpretation of these quantities.

Beyond just the porosity and permeability, various logging measurements allow the interpretation of what kinds of fluids are in the pores—oil, gas, brine. In addition, the logging measurements are used to determine mechanical properties of the formations. These mechanical properties determine what kind of enhanced recovery methods may be used (tertiary recovery) and what damage to the formation (such as erosion) is to be expected during oil and gas production.

The types of instruments used in well logging are quite broad. The first logging measurements consisted of basic electrical resistivity logs and spontaneous potential (SP) logs, introduced by the Schlumberger brothers in the 1920s. Tools later became available to estimate porosity via sonic velocity and nuclear measurements. Tools are now more specialised and better able to resolve fine details in the formation. Radiofrequency transmission and coupling techniques are used to determine electrical conductivity of fluid (brine is more conductive than oil or gas). Sonic transmission characteristics (pressure waves) determine mechanical integrity. Nuclear magnetic resonance (NMR) can determine the properties of the hydrogen atoms in the pores (surface tension, etc.). Nuclear scattering (radiation scattering), spectrometry and absorption measurements can determine density and elemental analysis or composition. High resolution electrical or acoustical imaging logs are used to visualize the formation, compute formation dip, and analyse thinly-bedded and fractured reservoirs.

In addition to sensor-based measurements above, robotic equipment can sample formation fluids which may then be brought to the surface for laboratory examination. Also, controlled flow measurements can be used to determine in situ viscosity, water and gas cut (percentage), and other fluid and production parameters.

Geological Logs

Geological logs use data collected at the surface, rather than by downhole instruments. The geological logs include *drilling time logs*, *core logs*, *sample logs*, and *mud logs*. Mud logs have become the oil industry standard.

Drilling time logs record the time required to drill a given thickness of rock formation. A change in the drilling rate or penetration rate usually means a change in the type of rock penetrated by the bit. The drilling time is expressed as minutes per foot, while the rate of penetration is usually expressed as feet per hour. Therefore, drilling time is the inverse of penetration rate.

Sample logs are made by examining cuttings, which are bits of rock circulated to the surface by the drilling mud in rotary drilling. The cuttings have travelled up the wellbore suspended in the drilling fluid or mud which was pumped into the wellbore via the drill string/pipe and they return to the surface via the annulus, then to the shale shakers via the flow line. Cuttings are then separated from the drilling fluid as they move across the shale shakers and are sampled at regular depth intervals. These rock samples are analysed and described by the wellsite geologist or mudlogger.

Mud logs are prepared by a mud logging company contracted by the operating company. One parameter a typical mud log displays is the formation gas (gas units or ppm). "The gas recorder usually is scaled in terms of arbitrary gas units, which are defined differently by the various gas-detector manufactures. In practice, significance is placed only on relative changes in the gas concentrations detected." The current industry standard mud log normally includes real-time drilling parameters such as rate of penetration (ROP), lithology, gas hydrocarbons, flow line temperature (temperature of the drilling fluid) and chlorides but may also include mud weight, estimated pore pressure and corrected d-exponent (corrected drilling exponent) for a pressure pack log. Other information that is normally notated on a mud log include lithology descriptions, directional data (deviation surveys), weight on bit, rotary speed, pump pressure, pump rate, viscosity, drill bit info, casing shoe depths, formation tops, mud pump info, to name just a few.

Wireline Log

A continuous measurement of formation properties with electrically powered instruments to infer properties and make decisions about drilling and production operations. The record of the measurements, typically a long strip of paper, is also called a log. Measurements include electrical properties (resistivity at various frequencies), sonic properties, active and passive nuclear measurements, dimensional measurements of the wellbore, formation fluid sampling, formation pressure measurement, wireline-conveyed sidewall coring tools, and others. In wireline measurements, the logging tool (or probe) is lowered into the open wellbore on a multiple conductor, contra-helically armoured wireline. Once lowered to the bottom of the interval of interest, the measurements are taken on the way out of the wellbore. This is done in an attempt to maintain tension on the cable (which stretches) as constant as possible for depth correlation purposes. (The exception to this practice is in certain hostile environments in which the tool electronics might not survive the temperatures on bottom for the amount of time it takes to lower the tool and then record measurements while pulling the tool up the hole. In this case, "down log" measurements might actually be conducted on the way into the well, and repeated on the way out if possible.) Most wireline measurements are recorded continuously even though the probe is moving. Certain fluid sampling and pressure-measuring tools require that the probe be stopped, increasing the chance that the probe or the cable might become stuck. LWD tools take measurements in much the same way as wireline-logging tools, except that the measurements are taken by a self-contained tool near the bottom of the bottomhole assembly and are recorded downward (as the well is deepened) rather than upward from the bottom of the hole (as wireline logs are recorded).

Memory Log

This method of data acquisition involves recording the sensor data into a down hole memory, rather than transmitting "Real Time" to surface. There are some advantages and disadvantages to this memory option.

- The tools can be conveyed into wells where the trajectory is deviated or extended beyond the reach of conventional Electric Wireline cables. This can involve a combination of weight to strength ratio of the electric cable over this extended reach. In such cases the memory tools can be conveyed on Pipe or Coil Tubing.
- The type of sensors are limited in comparison to those used on Electric Line, and tend to be focussed on the cased hole, production stage of the well. Although there are now developed some memory "Open Hole" compact formation evaluation tool combinations. These tools can be deployed and carried downhole concealed internally in drill pipe to protect them from damage while running in the hole, and then "Pumped" out the end at depth to initiate logging. Other basic open hole formation evaluation memory tools are available for use in "Commodity" markets on slickline to reduce costs and operating time.
- In cased hole operation there is normally a "Slick Line" intervention unit. This uses a solid mechanical wire (.82 -

.125 inches in OD), to manipulate or otherwise carry out operations in the well bore completion system. Memory operations are often carried out on this Slickline conveyance in preference to mobilising a full service Electric Wireline unit.

- Since the results are not known until returned to surface, any realtime well dynamic changes cannot be monitored real time. This limits the ability to modify or change the well down hole production conditions accurately during the memory logging by changing the surface production rates. Something that is often done in Eletric Line operations.
- Failure during recording is not known until the memory tools are retrieved. This loss of data can be a major issue on large offshore (expensive) locations. On land locations (e.g. South Texas, US) where there is what is called a "Commodity" Oil service sector, where logging often is without the rig infrastructure. This is less problematic, and logs are often run again without issue.

Information Use

In the oil industry, the well and mud logs are usually transferred in 'real time' to the operating company, which uses these logs to make operational decisions about the well, to correlate formation depths with surrounding wells, and to make interpretations about the quantity and quality of hydrocarbons present. Specialists involved in well log interpretation are called log analysts.

Gamma Ray Logging

Gamma ray logging is a method of measuring naturally occurring gamma radiation to characterise the rock or sediment in a borehole. It is sometimes used in mineral exploration and water-well drilling, but most commonly for formation evaluation in oil and gas well drilling. Different types of rock emit different amounts and different spectra of natural gamma radiation. In particular, shales usually emit more gamma rays than other sedimentary rocks, such as sandstone, gypsum, salt, coal, dolomite, or limestone because radioactive potassium is a common component in their clay content, and because the cation exchange capacity of clay causes them to adsorb uranium and thorium. This difference in radioactivity between shales and sandstones/carbonate rocks allows the gamma tool to distinguish between shales and non-shales.

The gamma ray log, like other types of well logging, is done simply by lowering an instrument down the hole and recording gamma radiation variation with depth. In the United States, the device most commonly records measurements at 1/2-foot intervals. Gamma radiation is usually recorded in API units, a measurement originated by the petroleum industry. Gamma logs are affected by the diametre of the borehole and the properties of the fluid filling the borehole, but because gamma logs are most often used in a qualitative way, corrections are usually not necessary. Three elements and their decay chains are responsible for the radiation emitted by rock: potassium, thorium and uranium. Shales often contain potassium as part of their clay content, and tend to absorb uranium and thorium as well. A common gamma-ray log records the total radiation, and cannot distinguish between the radioactive elements, while a spectral gamma ray log can.

For standard GR logs the value measured is calculated from thorium in ppm, Uranium in ppm and potassium in percent. GR API = 8 × Uranium concentration in ppm + 4 × thorium concentration in ppm + potassium concentration in percent. Due to the weight of uranium concentration in the calculation anomalous concentrations of uranium can cause clean sand reservoirs to appear shaley. Spectral Gamma ray is used to provide an individual reading for each element so anomalies in concentration can be found and interpreted.

An advantage of the gamma log over some other types of well logs is that it works through the steel and cement walls of cased boreholes. Although concrete and steel absorb some of the gamma radiation, enough travels through the steel and cement to allow qualitative determinations. Sometimes non-shales also have elevated levels of gamma radiation. Sandstone can contain uranium mineralization, potassium feldspar, clay filling, or rock fragments that cause it to have higher-than usual gamma readings. Coal and dolomite may contain absorbed uranium. Evaporite deposits may contain potassium minerals such as carnallite. When this is the case, spectral gamma ray logging can be done to identify these anomalies.

Spectral Logging

Some specialised gamma radiation logging distinguishes the three component decay chains (potassium, uranium, and thorium) by the wavelengths of their characteristic gamma emissions.

The characteristic gamma ray line that is associated with each component:

- Potassium : Gamma ray energy 1.46 MeV
- Thorium series: Gamma ray energy 2.62 MeV
- Uranium-Radium series: Gamma ray energy 1.76 MeV

Another example of the use of spectral gamma ray logs is to identify specific clay types, like Kaolinite or Illite. This can be used for environmental interpretation as Kaolinite forms from Feldspars in tropic soils by leaching of Potassium; and low Potassium readings may thus indicate paleosols. The identification of clay types is also useful for calculating the effective porosity of reservoir rock.

Use in Mineral Exploration

Gamma ray logs are also used in mineral exploration, especially exploration for phosphates, uranium, and potassium salts.

Resistivity Logging

Resistivity logging is a method of well logging that works by characterising the rock or sediment in a borehole by measuring its electrical resistivity. Resistivity is a fundamental material property which represents how strongly a material opposes the flow of electric current. In these logs, resistivity is measured using 4 electrical probes to eliminate the resistance of the contact leads. The log must run in holes containing electrically conductive mud or water.

Resistivity logging is sometimes used in mineral exploration and water-well drilling, but most commonly for formation evaluation in oil- and gas-well drilling. Most rock materials are essentially insulators, while their enclosed fluids are conductors. Hydrocarbon fluids are an exception, because they are almost infinitely resistive. When a formation is porous and contains salty water, the overall resistivity will be low. When the formation contains hydrocarbon, or contains very low porosity, its resistivity will be high. High resistivity values may indicate a hydrocarbon bearing formation.

Usually while drilling, drilling fluids invade the formation, changes in the resistivity are measured by the tool in the invaded zone. For this reason, several resistivity tools with different investigation lengths are used to measure the formation resistivity. If water based mud is used and oil is displaced, "deeper" resistivity logs (or those of the "virgin zone") will show lower conductivity than the invaded zone. If oil based mud is used and water is displaced, deeper logs will show higher conductivity than the invaded zone. This provides not only an indication of the fluids present, but also, at least quantitatively, whether the formation is permeable or not.

Use in Mineral Exploration

Resistivity logs are also used in mineral exploration, especially exploration for iron and potassium.

Density Logging

Density logging is a well logging tool determining rock bulk density along a wellbore. This is the overall density of a rock including solid matrix and the fluid enclosed in pores. Geologically, bulk density is a function of the density of the minerals forming a rock (i.e. matrix) and the enclosed volume of free fluids (porosity).

General Tool Design

Modern density logging tools, generally shares the same design principles independent of manufacturer. Tool consist of housing with electronics catridges and a sonde part. Sonde part has a measuring pad, where radiactive source (CS-137 nowadays) is inserted and a number of detectors (Long Space, Short Space, and a "Backscattering" detector).

Measuring pad is connected with a hydraulically operated caliper arm, which is used to measure borehole diametre, and on the same time apply pressure to the pad to ensure it's contact with the borehole. Density measurements are extremely sensitive to shape of the borehole - readings in large washouts are irrelevant due to very shallow depth of investigation of density tools.

Principle

A radioactive source applied to the hole wall emits medium-energy gamma rays into the formation so these gamma rays may be thought of as high velocity particles which collide with the electrons in the formation. At each collision the gamma ray loses some of its energy to the electron, and then continues with diminished energy. This type of interaction is known as Compton scattering. The scattered gamma rays reaching the detector, at the fixed station from the source, are counted as an indication of formation density.

The number of Compton scattering collisions is related directly to the number of the electron density of the formation. Consequently, the electron density determines the response of the density tool. Electron density is related through equation

$$\rho\dot{\epsilon} = \frac{2\rho B \sum Z_{1i}}{M}$$

where $\sum Z_{1i}$ is the atomic numbers of all the atoms making up the molecules in the compound, and M is them molecular weight of the compound.

Inferring Porosity from Bulk Density

Assuming that the measured bulk density (ρ_{bulk}) only depends on matrix density (ρ_{matrix}) and fluid density (ρ_{fluid}), and that these values are known along the wellbore, porosity (φ) can be inferred by the formula

$$\phi = \frac{\rho_{matrix} - \rho_{bulk}}{\rho_{matrix} - \rho_{fluid}}$$

Common values of matrix density ρ_{matrix} (in g/cm^3) are:

- Quartz sand - 2.65
- Limestone - 2.71
- Dolomite - 2.87

Density of clay minerals is highly variable, and depends on depositional environment, overburden pressure, type of clay mineral and many other factors. It can vary from 2.1 (Montmorillonite) to 2.76 (Chlorite). A fluid bulk density ρ_{fluid} of 1 g/cm^3 is appropriate where the water is fresh; highly saline water has a slightly higher density. For flushed gas or oil reservoirs, even lower ρ_{fluid} values should be assumed depending on the hydrocarbon density and residual saturation. In some applications hydrocarbons are indicated by the presence of abnormally high log porosities.

Sonic Logging

Sonic logging shows a formation's interval transit time, designated *Dt*. It is a measure of a formation's capacity to transmit sound waves. Geologically, this capacity varies with lithology and rock textures, notably porosity. Quantitatively, the sonic log is used to evaluate porosity in liquid filled pores. The sonic tool is only capable of measuring travel time. Many relationships between travel time and porosity have been proposed, the most commonly accepted is the Wyllie time average equation.

The equation basically holds that the total travel time recorded on the log is the sum of the time the sonic wave spends travelling the solid part of the rock, called the rock matrix and the time spent travelling through the fluids in the pores.

Use in Mineral Exploration

Sonic logs are also used in mineral exploration, especially exploration for iron, hydrocarbons and potassium.

Caliper Log

A caliper log is a set of measurements of the size and shape of a bore hole commonly made when drilling oil and gas wells. This can be an important indicator of cave ins or shale swelling in the bore hole.

The caliper tool measures the variation in bore hole diametre as it is withdrawn from the bottom of the hole. It is constructed with two or more articulated arms that push against the bore hole wall to take measurements. The arms show variable movements of the cursor by measuring electrical resistance, creating electrical variation. The variation in output is translated into changes of diametre after a simple calibration. The caliper log is printed as a continuous series of values of hole diametre with depth.

Known challenges with caliper logging include borehole spiralling. The position of the drill bit may precess as it drills, leading to spiralling shapes in the wellbore wall, as if the hole had been drilled by a screw. If the arms of the caliper log follow the grooves of the spiral, it will report too high an average diametre. Moving in and out of the grooves, the caliper will give erratic or periodically varying readings.

Mud Logging

Mud logging, also known as hydrocarbon well logging, is the creation of a detailed record (well log) of a borehole by examining the bits of rock or sediment brought to the surface by the circulating drilling medium (most commonly mud). Mud logging is usually performed by a third-party mud logging company. This provides well owners and producers with information about the lithology and fluid content of the borehole while drilling. Historically it is the earliest type of well log. Under some circumstances compressed air is employed as a circulating fluid, rather than mud. Although most commonly used in petroleum exploration, mud logging is also sometimes used when drilling water wells and in other mineral exploration, where drilling fluid is the circulating medium used to lift cuttings out of the hole. In hydrocarbon exploration, hydrocarbon surface gas detectors record the level of natural gas brought up in the mud. A mobile laboratory is situated by the mud logging company near the drilling rig or on deck of an offshore drilling rig, or on a drill ship.

The Service Contracted

A mud logging technician in the modern oil field drilling operation determines positions of hydrocarbons with respect to depth, identifies downhole lithology, monitors natural gas entering the drilling mud stream, and draws well logs for use by oil company geologists. Rock cuttings circulated to the surface in drilling mud are sampled and analysed.

The mud logging company is normally contracted by the oil company (or operator). They then organise this information in the form of a graphic log, showing the data charted on a representation of the wellbore.

The oil industry representative (or "company man") (also with the tool pusher, and geologist) provides mud loggers their instruction. The mud logging company is contracted specifically as to when to start well-logging activity and what services to provide. Mud logging may begin on the first day of drilling, known as the "spud in" date but is more likely at some later time (and depth) determined by the oil industry geologist's research. The mud logger may also possess logs from wells drilled in the surrounding area. This information (known as "offset data") can provide valuable clues as to the characteristics of the particular geo-strata that the rig crew is about to drill through.

Mud loggers connect various sensors to the drilling apparatus and install specialised equipment to monitor or "log" drill activity. This can be physically and mentally challenging, especially when having to be done during drilling activity. Much of the equipment will require precise calibration or alignment by the mud logger to provide accurate readings. Mud logging technicians observe and interpret the indicators in the mud returns during the drilling process, and at regular intervals log properties such as drilling rate, mud weight, flowline temperature, oil indicators, pump pressure, pump rate, lithology (rock type) of the drilled cuttings, and other data. Mud logging requires a good deal of diligence and attention. Sampling the drilled cuttings must be performed at predetermined intervals, and can be difficult during rapid drilling.

Another important task of the mud logger is to monitor gas levels (and types) and notify other personnel on the rig when gas levels may be reaching dangerous levels, so appropriate steps can be taken to avoid a dangerous well blowout condition. Because of the lag time between drilling and the time required for the mud and cuttings to return to the surface, a modern augmentation has come into use: Measurement while drilling. The MWD technician, often a separate service company employee, logs data in a similar manner but whose data is different in source and content. Most of the data logged by an MWD technician comes from expensive and complex, sometimes electronic, tools that are downhole installed at or near the drill bit.

Scope

Mud logging includes observation and microscopic examination of drill cuttings (formation rock chips), and evaluation of gas hydrocarbon and its constituents, basic chemical and mechanical parameters of drilling fluid or drilling mud (such as chlorides and temperature), as well as compiling other information about the drilling parameters. Then data is plotted on a graphic log called a mud log. Example1, Example2.

Other real-time drilling parameters that may be compiled include, but are not limited to; rate of penetration (ROP) of the bit (sometimes called the drill rate), pump rate (quantity of fluid being pumped), pump pressure, weight on bit, drill string weight, rotary speed, rotary torque, RPM (Revolutions Per Minute), SPM (Strokes Per Minute) mud volumes, mud weight and mud viscosity. This information is usually obtained by attaching *monitoring devices* to the drilling rig's equipment with a few exceptions such as the mud weight and mud viscosity which are measured by the derrickhand or the mud engineer.

Rate of drilling is affected by the pressure of the column of mud in the borehole and its relative counterbalance to the internal pore pressures of the encountered rock. A rock pressure greater than the mud fluid will tend to cause rock fragments to spall as it is cut and can increase the drilling rate. "D-exponents" are mathematical trend lines which estimate this internal pressure. Thus both visual evidence of spalling and mathematical plotting assist in formulating recommendations for optimum drilling mud densities for both safety (blowout prevention) and economics. (Faster drilling is generally preferred.)

Mud logging is often written as a single word "mudlogging". The finished product can be called a "mud log" or "mudlog". The occupational description is "mud logger" or "mudlogger". In most cases, the two word usage seems to be more common. The mud log provides a reliable *time log* of drilled formations.

Details of the Mud Log

- The *rate of penetration* is represented by the black line on the left side of the log. The farther to the left that the line goes, the faster the rate of penetration. On this mud log, ROP is measured in feet per hour but on some older, hand drawn mud logs, it is measured in minutes per foot.
- The *porosity* is represented by the blue line farthest to the left of the log. It indicates the pore space within the rock structure. An analogy would be the holes in a sponge. The oil and gas resides within this pore space. Notice how far to the left the porosity goes where all the sand (in yellow) is. This indicates that the sand has good porosity. Porosity is not a direct or physical measurement of the pore space but rather an extrapolation from other drilling parameters and therefore not always reliable.

Figure: *Sample of drill cuttings of shale while drilling an oil well in Louisiana. For reference, the sand grain and red shale are approximately 2 mm. in dia.*

- The *lithology* is represented by the cyan, gray/black and yellow blocks of colour. Cyan = lime, gray/black = shale and yellow = sand. More yellow represents more sand identified at that depth. The lithology is measured as percentage of the total sample, as visually inspected under a microscope, normally at 10x magnification. These are but a fraction of the different types of formations that might be encountered. (Colour coding is not necessarily standardized among different mud logging companies, though the symbol representation for each are very similar.) In you can see a sample of cuttings under a microscope at 10x magnification after they have been washed off. Some of the larger shale and lime fragments are separated from this sample by running it through sieves and must be considered when estimating percentages. Also, this image view is only a fragment of the total sample and some of the sand at the bottom of the tray can not be seen and must also be considered in the total estimation. With that in mind this sample would be considered to be about 90% shale, 5% sand and 5% lime (In 5% increments).
- The *gas* is represented by the green line and is measured in units of ppm (parts per million) as the quantity of total gas, but does not represent the actual quantity of oil or gas the reservoir contains. In the squared-off dash-dot lines just to the right of

the sand (in yellow) and left of the gas (in green) represents the heavier hydrocarbons detected. Cyan = C_2 (ethane), purple = C_3 (propane) and blue = C_4 (butane). Detecting and analysing these heavy gases help to determine the type of oil or gas the formation contains.

Logging While Drilling

Logging While Drilling (LWD) is a technique of conveying well logging tools into the well borehole downhole as part of the bottom hole assembly (BHA). LWD tools work with its Measurement While Drilling (MWD) system to transmit partial or complete measurement results to the surface via typically a drilling mud pulser or other improved techniques, while LWD tools are still in the borehole, which is called "Real Time Data". Complete measurement results can be downloaded from LWD tools after they are pulled out of hole, which is called "Memory Data".

LWD technology was developed originally as an enhancement to the earlier MWD technology to completely or partially replace wireline logging operation. With the improvement of the technology in the past decades, LWD is now widely used for drilling (including geosteering), formation evaluation (especially for real time and high angle wells).

Available LWD Measurements

LWD technology was originally developed to partially or completely replace wireline logging. Over the years, majority of the measurements have been made available in LWD. Certain new measurements are also development in LWD only. The following is an incomplete list of available measurement in LWD technology.

- Natural Gamma Ray (GR)
  - Total Gamma Ray
  - Spectral Gamma Ray
  - Azimuthal Gamma Ray
  - Gamma ray close to drill bit.
- Density and Photoelectric Index
- Neutron Porosity
- Borehole Caliper
  - Ultra sonic azimuthal caliper.
  - Density Caliper
- Resistivity (ohm-m)
  - Attenuation and phase shift resistivities at different transmitter spacings and frequencies.

- Resistivity at the drill bit.
- Deep directional resistivitiesjhhi

- Sonic
  - Compressional Slowness(Dtc)
  - Shear Slowness (Dts)
- Borehole Images
  - Density Borehole Image
  - Resistivity Borehole Image
- Formation Tester and Sampler
  - Formation Pressure
  - Formation Fluid Sample
- Nuclear Magnetic Resonance (NMR)
- Seismic While Drilling (SWD)
  - Drillbit-SWD

Steam Injection (Oil Industry)

Steam injection is an increasingly common method of extracting heavy oil. It is considered an enhanced oil recovery (EOR) method and is the main type of thermal stimulation of oil reservoirs. There are several different forms of the technology, with the two main ones being Cyclic Steam Stimulation and Steam Flooding. Both are most commonly applied to oil reservoirs, which are relatively shallow and which contain crude oils which are very viscous at the temperature of the native underground formation. Steam injection is widely used in the San Joaquin Valley of California (USA), the Lake Maracaibo area of Venezuela, and the oil sands of northern Alberta (Canada).

Cyclic Steam Stimulation

This method, also known as the Huff and Puff method, consists of 3 stages: injection, soaking, and production. Steam is first injected into a well for a certain amount of time to heat the oil in the surrounding reservoir to a temperature at which it flows. After it is decided enough steam has been injected, the steam is usually left to "soak" for some time after (typically not more than a few days). Then oil is produced out of the same well, at first by natural flow (since the steam injection will have increased the reservoir pressure) and then by artificial lift. Production will decrease as the oil cools down, and once production reaches an economically determined level the steps are repeated again.

The process can be quite effective, especially in the first few cycles. However, it is typically only able to recover approximately 20% of the Original Oil in Place (OOIP), compared to steam flooding, which has been reported to recover over 50% of OOIP. It is quite common for wells to be produced in the cyclic steam manner for a few cycles before being put on a steam flooding regime with other wells.

The mechanism was accidentally discovered by Shell while it was doing a steam flood in Venezuela and one of its steam injectors blew out and ended up producing oil at much higher rates than a conventional production well in a similar environment.

Steam Flooding

In a steam flood, sometimes known as a steam drive, some wells are used as steam injection wells and other wells are used for oil production. Two mechanisms are at work to improve the amount of oil recovered. The first is to heat the oil to higher temperatures and to thereby decrease its viscosity so that it more easily flows through the formation toward the producing wells. A second mechanism is the physical displacement employing in a manner similar to water flooding, in which oil is meant to be pushed to the production wells. While more steam is needed for this method than for the cyclic method, it is typically more effective at recovering a larger portion of the oil.

A form of steam flooding that has become popular in the Alberta oil sands is steam assisted gravity drainage (SAGD), in which two horizontal wells are drilled, one a few metres above the other, and steam is injected into the upper one. The intent is to reduce the viscosity of the bitumen to the point where gravity will pull it down into the producing well.

In 2011 Laricina Energy combined solvent injection with steam injection in a process called solvent cyclic steam-assisted gravity drainage (SC-SAGD). Laricina claims that combining solvents with steam reduces the overall steam oil ratio for recovery by 30%.

Water Injection (Oil Production)

Water injection refers to the method in oil industry where water is injected back into the reservoir, usually to increase pressure and thereby stimulate production. Water injection wells can be found both on- and offshore, to increase oil recovery from an existing reservoir. The first use of water injection to increase production from failing oil wells was in the states of New York and Pennsylvania in the early 1930s.

Water is injected (1) to support pressure of the reservoir (also known as voidage replacement), and (2) to sweep or displace oil from the reservoir, and push it towards a well. Normally only 30% of the oil in a reservoir can be extracted, but water injection increases that percentage (known as the recovery factor) and maintains the production rate of a reservoir over a longer period of time.

Sources of Injected Water

Any and every source of bulk water can be, and has been, used for injection. The following sources of water are used for recovery of oil: Produced water is often used as an injection fluid. This reduces the potential of causing formation damage due to incompatible fluids, although the risk of scaling or corrosion in injection flowlines or tubing remains. Also, the produced water, being contaminated with hydrocarbons and solids, must be disposed of in some manner, and disposal to sea or river will require a certain level of clean-up of the water stream first. However, the processing required to render produced water fit for reinjection may be equally costly.

As the volumes of water being produced are never sufficient to replace all the production volumes (oil and gas, in addition to water), additional "make-up" water must be provided. Mixing waters from different sources exacerbates the risk of scaling. Seawater is obviously the most convenient source for offshore production facilities, and it may be pumped inshore for use in land fields. Where possible, the water intake is placed at sufficient depth to reduce the concentration of algae; however, filtering, deoxygenation and biociding is generally required.

Aquifer water from water-bearing formations other than the oil reservoir, but in the same structure, has the advantage of purity where available. River water will always require filtering and biociding before injection.

Filters

The filters must clean the water and remove any impurities, such as shells and algae. Typical filtration is to 2 micrometres, but really depends on reservoir requirements. The filters are so fine so as not to block the pores of the reservoir. Sand filters are a common used filtration technology to remove solid impurities from the water. The sand filter has different beds with various sizes of sand granules. The sea water traverses the first, finest, layer of sand down to the coarsest and to clean the filter, the process is inverted. After the water is filtered it continues on to fill the de-oxygenation tower. Sand filters are bulky, heavy, have some spill over of sand particles and require chemicals to enhance water quality.

A more sophisticated approach is to use automatic selfcleaning backflushable screen filters (suction scanning) because these do not have the disadvantages sand filters have.

The importance of proper water treatment is often underestimated by oil companies and engineering companies. *Especially with river-, and seawater, intake water quality can vary tremendously (algea blooming in spring time, storms and current stirring up sediments from the seafloor) which will have significant impact on the performance of the water treatment facilities.* If not addressed correctly, water injection may not be successful. This results in poor water quality, clogging of the reservoir and loss of oil production.

De-Oxygenation

Oxygen must be removed from the water because it promotes corrosion and growth of certain bacteria. Bacterial growth in the reservoir can produce toxic hydrogen sulfide, a source of serious production problems, and block the pores in the rock.

A deoxygenation tower brings the injection water into contact with a dry gas stream (gas is always readily available in the oilfield). The filtered water drops into the de-oxygenation tower, splashing onto a series of trays, causing dissolved oxygen to be lost to the gas stream.

An alternative method, also used as a backup to deoxygenation towers, is to add an oxygen scavenging agent such as sodium bisulfite.

Water Injection Pumps

The high pressure, high flow water injection pumps are placed near to the de-oxygenation tower and boosting pumps. They fill the bottom of the reservoir with the filtered water to push the oil towards the wells like a piston. The result of the injection is not quick, it needs time.

Water injection is used to prevent low pressure in the reservoir. The water replaces the oil which has been taken, keeping the production rate and the pressure the same over the long term.

Upstream (Petroleum Industry)

The oil and gas industry is usually divided into three major components: Upstream, midstream and downstream, though midstream operations are usually included in the downstream category.

The upstream oil sector is a term commonly used to refer to the searching for and the recovery and production of crude oil and natural gas. The upstream oil sector is also known as the *exploration and production (E&P) sector*.

The upstream sector includes the searching for potential underground or underwater oil and gas fields, drilling of exploratory wells, and subsequently operating the wells that recover and bring the crude oil and/or raw natural gas to the surface.

Downstream (Petroleum Industry)

The petroleum industry is usually divided into three major components: Upstream, midstream and downstream. Midstream operations are usually included in the downstream category.

The downstream oil sector is a term commonly used to refer to the refining of crude oil, and the selling and distribution of natural gas and products derived from crude oil. Such products include liquified petroleum gas (LPG), gasoline or petrol, jet fuel, diesel oil, other fuel oils, asphalt and petroleum coke.

The downstream sector includes oil refineries, petrochemical plants, petroleum product distribution, retail outlets and natural gas distribution companies. The downstream industry touches consumers through thousands of products such as petrol, diesel, jet fuel, heating oil, asphalt, lubricants, synthetic rubber, plastics, fertilizers, antifreeze, pesticides, pharmaceuticals, natural gas, and propane.

Byproduct Sulfur

Crude oil is a mixture of many varieties of hydrocarbons, and may contain sulfurous compounds. The refining process converts most of that sulfur into hydrogen sulfide. Raw natural gas also may contain gaseous hydrogen sulphide and sulfur-containing mercaptans, which are removed in natural gas processing plants before the gas is distributed to consumers.

The hydrogen sulfide removed in the refining and processing of crude oil and natural gas is subsequently converted into byproduct elemental sulfur. In fact, the vast majority of the 64,000,000 metric tons of sulfur produced worldwide in 2005 was byproduct sulphur from refineries and natural gas processing plants.

Reservoir Simulation

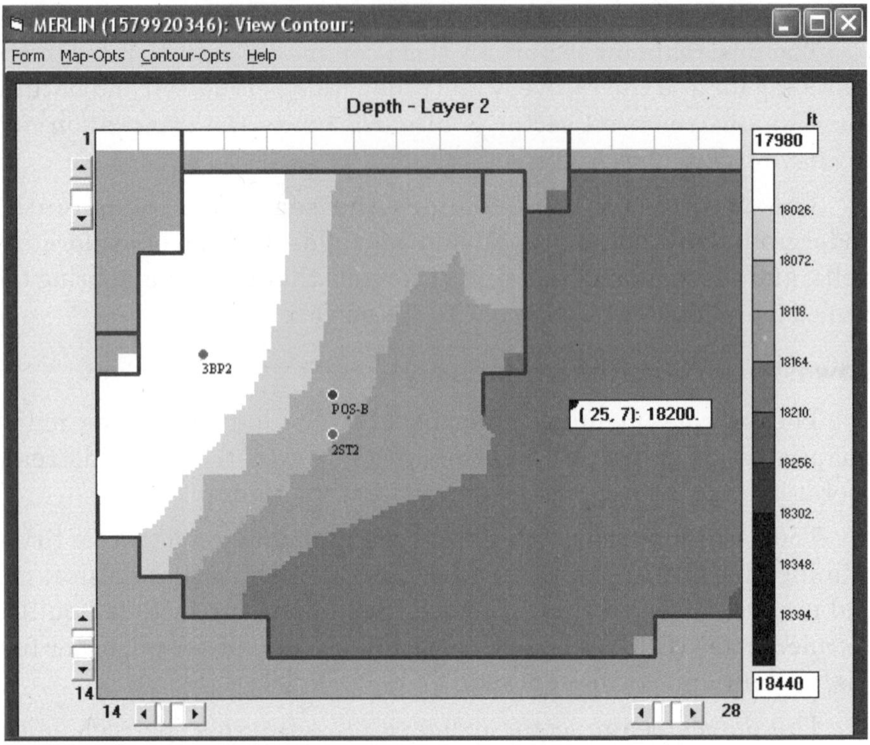

Figure: *A simulated depth map of the geology in a full field model from the Merlin finite difference simulator*

Reservoir simulation is an area of reservoir engineering in which computer models are used to predict the flow of fluids (typically, oil, water, and gas) through porous media.

Uses

Reservoir simulation models are used by oil and gas companies in the development of new fields. Also, models are used in developed fields where production forecasts are needed to help make investment decisions. As building and maintaining a robust, reliable model of a field is often time-consuming and expensive, models are typically only constructed where large investment decisions are at stake. Improvements in simulation software have lowered the time to develop a model. Also, models can be run on personal computers rather than more expensive workstations.

For new fields, models may help development by identifying the number of wells required, the optimal completion of wells, the present and future needs for artificial lift, and the expected production of oil, water and gas. For ongoing reservoir management, models may help in improved oil recovery by hydraulic fracturing. Highly deviated or horizontal wells can also be represented. Specialised software may be used in the design of hydraulic fracturing, then the improvements in productivity can be included in the field model. Also, future improvement in oil recovery with pressure maintenance by re-injection of produced gas or by water injection into an aquifer can be evaluated. Water flooding resulting in the improved displacement of oil is commonly evaluated using reservoir simulation.

The application of enhanced oil recovery (EOR) processes requires that the field possesses the necessary characteristics to make application successful. Model studies can assist in this evaluation. EOR processes include miscible displacement by natural gas, CO_2, or nitrogen and chemical flooding (polymer, alkaline, surfactant, or a combination of these). Special features in simulation software is needed to represent these processes. In some miscible applications, the "smearing" of the flood front, also called numerical dispersion, may be a problem.

Reservoir simulation is used extensively to identify opportunities to increase oil production in heavy oil deposits. Oil recovery is improved by lowering the oil viscosity by injecting steam or hot water. Typical processes are steam soaks (steam is injected, then oil produced from the same well) and steam flooding (separate steam injectors and oil producers). These processes require simulators with special features to account for heat transfer to the fluids present and the formation, the subsequent property changes and heat losses outside of the formation.

A recent application of reservoir simulation is the modelling of coalbed methane (CBM) production. This application requires a specialised CBM simulator. In addition to the normal fractured (fissured) formation data, CBM simulation requires gas content data values at initial pressure, sorption isotherms, diffusion coefficient, and parameters to estimate the changes in absolute permeability as a function of pore-pressure depletion and gas desorption.

Fundamentals

Traditional finite difference simulators dominate both theoretical and practical work in reservoir simulation. Conventional FD simulation is underpinned by three physical concepts: conservation of mass, isothermal fluid phase behaviour, and the Darcy approximation of fluid flow through porous media. Thermal simulators (most commonly used for heavy oil applications) add conservation of energy to this list, allowing temperatures to change within the reservoir.

Numerical techniques and approaches that are common in modern simulators:

- Most modern FD simulation programs allow for construction of 3-D representations for use in either full-field or single-well models. 2-D approximations are also used in various conceptual models, such as cross-sections and 2-D radial grid models.
- Theoretically, finite difference models permit discretization of the reservoir using both structured and more complex unstructured grids to accurately represent the geometry of the reservoir. Local grid refinements (a finer grid embedded inside of a coarse grid) are also a feature provided by many simulators to more accurately represent the near wellbore multi-phase flow affects. This "refined meshing" near wellbores is extremely important when analysing issues such as water and gas coning in reservoirs.
- Representation of faults and their transmissibilities are advanced features provided in many simulators. In these models, inter-cell flow transmissibilities must be computed for non-adjacent layers outside of conventional neighbour-to-neighbour connections.
- Natural fracture simulation (known as dual-porosity and dual-permeability) is an advanced feature which model hydrocarbons in tight matrix blocks. Flow occurs from the tight matrix blocks to the more permeable fracture networks that surround the blocks, and to the wells.
- A black oil simulator does not consider changes in composition of the hydrocarbons as the field is produced. The compositional model, is a more complex model, where the PVT properties of oil and gas phases have been fitted to an equation of state (EOS), as a mixture of components. The simulator then uses the fitted EOS equation to dynamically track the movement of both phases and components in field.

The simulation model computes the saturation change of three phases (oil, water and gas) and pressure of each phase in each cell at each time step. As a result of declining pressure as in a reservoir depletion study, gas will be liberated from the oil. If pressures increase as a result of water or gas injection, the gas is re-dissolved into the oil phase.

A simulation project of a developed field, usually requires "history matching" where historical field production and pressures are compared to calculated values. In recent years optimisation tools such as MEPO has helped to accelerate this process, as well as improve the quality of the match obtained. The model's parameters are adjusted until a reasonable match is achieved on a field basis and usually for all wells. Commonly, producing water cuts or water-oil ratios and gas-oil ratios are matched.

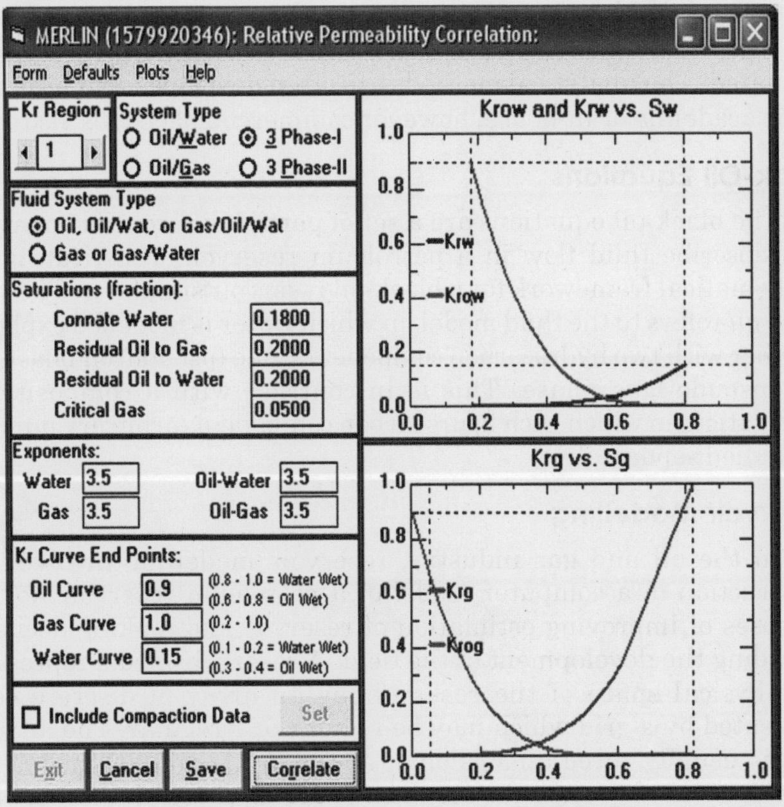

Figure: *Correlating relative permeability*

Other types of simulators include finite element and streamline.

Other Engineering Approaches

Without FD models, recovery estimates and oil rates can also be calculated using numerous analytical techniques which include material balance equations (including Havlena-Odeh and Tarner method), fractional flow curve methods (1-D displacement by Buckley-Leverett, Deitz method for inclined structures, coning models), sweep efficiency estimation techniques for water floods and decline curve analysis. These methods were developed and used prior to traditional or "conventional" simulations tools as computationally inexpensive models based on simple homogeneous reservoir description. Analytical methods generally cannot capture all the details of the given reservoir or process, but are typically numerically fast and at times, sufficiently reliable. In modern reservoir engineering, they are generally used as screening or preliminary evaluation tools. Analytical methods are especially suitable for potential assets evaluation when the data are limited and the time is critical, or for broad studies as a pre-screening tool if a large number of processes and / or technologies are to be evaluated. The analytical methods are often developed and promoted in the academia or in-house, however commercial packages also exist.

Black-Oil Equations

The black-oil equations are a set of partial differential equations that describe fluid flow in a petroleum reservoir, constituting the mathematical framework for a black-oil reservoir simulator. The term *black-oil* refers to the fluid model, in which water is modelled explicitly together with two hydrocarbon components, one (pseudo) oil phase and one (pseudo-)gas phase. This is in contrast with a compositional formulation, in which each hydrocarbon component (arbitrary number) is handled separately

Reservoir Modelling

In the oil and gas industry, reservoir modelling involves the construction of a computer model of a petroleum reservoir, for the purposes of improving estimation of reserves and making decisions regarding the development of the field. A reservoir model represents the physical space of the reservoir by an array of discrete cells, delineated by a grid which may be regular or irregular. The array of cells is usually three dimensional, although 1D and 2D models are sometimes used. Values for attributes such as porosity, permeability and water saturation are associated with each cell. The value of each attribute is implicitly deemed to apply uniformly throughout the volume of the reservoir represented by the cell.

Types of Reservoir Model

Reservoir models typically fall into two categories:

- Geological models are created by geologists and geophysicists and aim to provide a static description of the reservoir, prior to production.
- Reservoir simulation models are created by reservoir engineers and use finite difference methods to simulate the flow of fluids within the reservoir, over its production lifetime.

Sometimes a single "shared earth model" is used for both purposes. More commonly, a geological model is constructed at a relatively high (fine) resolution. A coarser grid for the reservoir simulation model is constructed, with perhaps two orders of magnitude fewer cells. Effective values of attributes for the simulation model are then derived from the geological model by a process of "upscaling". Uncertainty in the true values of the reservoir properties is sometimes investigated by constructing several different realisations of the sets of attribute values. The behaviour of the resulting simulation models can then indicate the associated level of economic uncertainty.

The phrase "reservoir characterisation" is sometimes used to refer to reservoir modelling activities up to the point when a simulation model is ready to simulate the flow of fluids. Commercially available software is used in the construction, simulation and analysis of the reservoir models.

Seismic to Simulation

Seismic to Simulation is the process and associated techniques used to develop highly accurate static and dynamic 3D models of hydrocarbon reservoirs for use in predicting future production, placing additional wells, and evaluating alternative reservoir management scenarios. The process is successful if the model accurately reflects the original well logs, seismic data and production history.

Reservoir models are constructed to gain a better understanding of the subsurface that leads to informed well placement, reserves estimation and production planning. Models are based on measurements taken in the field, including well logs, seismic surveys, and production history.

Seismic to simulation enables the quantitative integration of all field data into an updateable reservoir model built by a team of geologists, geophysicists, and engineers. Key techniques used in the process include integrated petrophysics and rock physics to determine the range of lithotypes and rock properties, geostatistical inversion to determine a set of plausible seismic-derived rock property models at sufficient vertical resolution and heterogeneity for flow simulation, stratigraphic grid transfer to accurately move seismic-derived data to the geologic model, and flow simulation for model validation and ranking to determine the model that best fits all the data.

Rock Physics and Petrophysics

The first step in seismic to simulation is establishing a relationship between petrophysical key rock properties and elastic properties of the rock. This is required in order to find common ground between the well logs and seismic data. Well logs are measured in depth and provide high resolution vertical data, but no insight into the inter-well space. Seismic are measured in time and provide great lateral detail but is quite limited in its vertical resolution. When correlated, well logs and seismic can be used to create a fine-scale 3D model of the subsurface.

Insight into the rock properties comes from a combination of basic geologic understanding and well bore measurements. Based on an understanding of how the area was formed over time, geologists can predict the types of rock likely to be present and how rapidly they vary spatially. Well log and core measurements provide samples to verify and fine-tune that understanding.

Seismic data is used by petrophysicists to identify the tops of various lithotypes and the distribution of rock properties in the inter-well space using seismic inversion attributes such as impedance. Seismic surveys measure acoustic impedance contrasts between rock layers. As different geologic structures are encountered, the sound wave reflects and refracts as a function of the impedance contrast between the layers. Acoustic impedance varies by rock type and can therefore be correlated to rock properties using rock physics relationships between the inversion attributes and petrophysical properties such as porosity, lithology, water saturation, and permeability.

Once well logs are properly conditioned and edited, a petrophysical rock model is generated that can be used to derive the effective elastic rock properties from fluid and mineral parameters as well as rock structure information. The model parameters are calibrated by comparison of the synthetic to the available elastic sonic logs. Calculations are performed following a number of rock physics algorithms including: Xu & White, Greenberg & Castagna, Gassmann, Gardner, modified upper and lower Hashin-Shtrikman, and Batzle & Wang.

When the petrophysical rock model is complete, a statistical database is created to describe the rock types and their known properties such as porosity and permeability. Lithotypes are described, along with their distinct elastic properties.

MCMC Geostatistical Inversion

In the next step of seismic to simulation, seismic inversion techniques combine well and seismic data to produce multiple equally plausible 3D models of the elastic properties of the reservoir. Seismic data is transformed to elastic property log(s) at every trace. Deterministic inversion techniques are used to provide a good overall view of the porosity over the field, and serve as a quality control check. To obtain greater detail needed for complex geology, additional stochastic inversion is then employed.

Geostatistical inversion procedures detect and delineate thin reservoirs otherwise poorly defined. Markov chain Monte Carlo (MCMC) based geostatistical inversion addresses the vertical scaling problem by creating seismic derived rock properties with vertical sampling compatible to geologic models. All field data is incorporated into the geostatistical inversion process through the use of probability distribution functions (PDFs). Each PDF describes a particular input data in geostatistical terms using histograms and variograms, which identify the odds of a given value at a specific place and the overall expected scale and texture based on geologic insight.

Once constructed, the PDFs are combined using Bayesian inference, resulting in a posterior PDF that conforms to everything that is known about the field. A weighting system is used within the algorithm, making the process more objective.

From the posterior PDF, realisations are generated using a Markov chain Monte Carlo algorithm. These realisations are statistically fair and produce models of high detail, accuracy and realism. Rock properties like porosity can be cosimulated from the elastic properties determined by the geostatistical inversion. This process is iterated until a best fit model is identified. Inversion parameters are tuned by running the inversion many times with and without well data. Without the well data, the inversions are running in blind-well mode. These blind-well mode inversions test the reliability of the constrained inversion and remove potential biased.

This statistical approach creates multiple, equi-probable models consistent with the seismic, wells, and geology. Geostatistical inversion simultaneously inverts for impedance and discrete properties types, and other petrophysical properties such as porosity can then be jointly cosimulated. The output volumes are at a sample rate consistent with the reservoir model because making synthetics of finely sampled models is the same as from well logs. Inversion properties are consistent with well log properties because the histograms used to generate the output rock properties from the inversion are based on well log values for those rock properties.

Uncertainty is quantified by using random seeds to generate slightly differing realisations, particularly for areas of interest. This process improves the understanding of uncertainty and risk within the model.

Stratigraphic Grid Transfer

Following geostatistical inversion and in preparation for history matching and flow simulation, the static model is re-gridded and up-scaled. The transfer simultaneously converts time to depth for the various properties and transfers them in 3D from the seismic grid to a corner-point grid. The relative locations of properties are preserved, ensuring data points in the seismic grid arrive in the correct stratigraphic layer in the corner point grid.

The static model built from seismic is typically orthogonal but flow simulators expect corner point grids. The corner point grid consists of cubes that are usually much coarser in the horizontal direction and each corner of the cube is arbitrarily defined to follow the major features in the grid. Converting directly from orthogonal to corner point can cause problems such as creating discontinuity in fluid flow.

An intermediate stratigraphic grid ensures that important structures are not misrepresented in the transfer. The stratigraphic grid has the same number of cells as the orthogonal seismic grid, but the boundaries are defined by stratigraphic surfaces and the cells follow the stratigraphic organisation. This is a stratigraphic representation of the seismic data using the seismic interpretation to define the layers. The stratigraphic grid model is then mapped to the corner point grid by adjusting the zones.

Using the porosity and permeability models and a saturation height function, initial saturation models are built. If volumetric calculations identify problems in the model, changes are made in the petrophysical model without causing the model to stray from the original input data. For example, sealing faults are added for greater compartmentalization.

Model Validation and Ranking

In the last step of seismic to simulation, flow simulation continues the integration process by bringing in the production history. This provides a further validation of the static model against history. A representative set of the model realisations from the geostatistical inversion are history matched against production data. If the properties in the model are realistic, simulated well bottom hole pressure behaviour should match historical (measured) well bottom hole pressure. Production flow rates and other engineering data should also match.

Based on the quality of the match, some models are eliminated. After the initial history match process, dynamic well parameters are adjusted as needed for each of the remaining models to improve the match. The final model represents the best match to original field measurements and production data and is then used in drilling decisions and production planning.

Seismic Inversion

Seismic inversion, in Geophysics (primarily Oil and Gas exploration/development), is the process of transforming seismic reflection data into a quantitative rock-property description of a reservoir. Seismic inversion may be pre-or post-stack, deterministic, random or geostatistical, and typically includes other reservoir measurements such as well logs and cores. Seismic surveys are routinely performed to gather information about the geology of an oil or gas field. These surveys record sound waves which have travelled through the layers of rock and fluid in the earth. The amplitude and frequency of these waves can be estimated so that any side-lobe and tuning effects introduced by the wavelet may be removed.

Seismic data may be inspected and interpreted on its own without inversion, but this does not provide the most detailed view of the subsurface and can be misleading under certain conditions. Because of its efficiency and quality, most oil and gas companies now use seismic inversion to increase the resolution and reliability of the data and to improve estimation of rock properties including porosity and net pay.

There are many different techniques used in seismic inversion. These can be roughly grouped into two categories: pre-stack or post-stack, and seismic resolution or well-log resolution. The combination of these categories yields four technical approaches to the inversion problem, and the selection of a specific technique depends on the desired objective and the characteristics of the subsurface rocks. Although the order presented reflects the advancement in inversion techniques over the past 20 years, each grouping still has valid uses in particular projects or as part of a larger workflow.

Wavelet Estimation

All modern seismic inversion methods require seismic data and a wavelet estimated from the data. Typically, a reflection coefficient series from a well within the boundaries of the seismic survey is used to estimate the wavelet phase and frequency. Accurate wavelet estimation is critical to the success of any seismic inversion. The inferred shape of the seismic wavelet may strongly influence the seismic inversion results and, thus, subsequent assessments of the reservoir quality.

Wavelet amplitude and phase spectra are estimated statistically from either the seismic data alone or from a combination of seismic data and well control using wells with available sonic and density curves. After the seismic wavelet is estimated, it is used to estimate seismic reflection coefficients in the seismic inversion.

When the estimated (constant) phase of the statistical wavelet is consistent with the final result, the wavelet estimation converges more quickly than when starting with a zero phase assumption. Minor edits and "stretch and squeeze" may be applied to the well to better align the events. Accurate wavelet estimation requires the accurate tie of the impedance log to the seismic. Errors in well tie can result in phase or frequency artifacts in the wavelet estimation. Once the wavelet is identified, seismic inversion computes a synthetic log for every seismic trace. To ensure quality, the inversion result is convolved with the wavelet to produce synthetic seismic traces which are compared to the original seismic.

Components of Inversion

Inversion includes both seismic data and well data, where well data serves to add the low frequency below the seismic band and to constrain the inversion. Well logs are first conditioned and edited to ensure there is a suitable relationship between impedance logs and the desired properties. The logs are then converted to time, filtered to approximate the seismic bandwidth and edited for borehole effects, balanced and classified by quality.

Seismic data is band-limited, reducing resolution and quality. To extend the frequency band available, low-frequency data is derived from log data, pre-stack depth or time migrated velocities and/or a regional gradient. High frequency can be derived from well control or geostatistical analysis. Initial inversions are often run with relaxed constraints, starting with the seismic and then adding limited-trend data from the wells. This provides a rough overview of the reservoir in an unbiased manner. It is critical at this point to evaluate the accuracy of the tie between the inversion results and the wells, and between the original seismic data and the derived synthetics. It is also important to ensure that the wavelet matches the phase and frequency of seismic data. Without a wavelet, the solution is not unique. Deterministic inversions address this problem by constraining the answer in some way, usually to well log data. Stochastic inversions address this problem by generating a range of plausible solutions, which can then be narrowed through testing for best fit against various measurements (including production data).

Post-Stack Seismic Resolution Inversion

An example of a post-stack seismic resolution inversion technique is the Constrained Sparse-Spike Inversion (CSSI). This assumes a limited number of reflection coefficients, with larger amplitude. The inversion results in acoustic impedance (AI), which is the product of rock density and p-wave velocity. Unlike seismic reflection data (which is an interface property) AI is a rock property. The model generated is of higher quality, and does not suffer from tuning and interference caused by the wavelet. CSSI transforms seismic data to a pseudo-acoustic impedance log at every trace. Acoustic impedance is used to produce more accurate and detailed structural and stratigraphic interpretations than can be obtained from seismic (or seismic attribute) interpretation. In many geological environments acoustic impedance has a strong relationship to petrophysical properties such as porosity, lithology, and fluid saturation.

A good (CSSI) algorithm will produce four high-quality acoustic impedance volumes from full or post-stack seismic data: full-bandwidth impedance, bandlimited Impedance, reflectivity model, and low-frequency component. Each of these components can be inspected for its contribution to the solution and to check the results for quality. To further adapt the algorithm mathematics to the behaviour of real rocks in the subsurface, some CSSI algorithms use a mixed-norm approach and allow a weighting factor between minimising the sparsity of the solution and minimising the misfit of the residual traces.

Pre-Stack Seismic Resolution Inversion

Pre-stack inversion is often used when post-stack inversion fails to sufficiently differentiate geologic features with similar P-impedance signatures. Simultaneous inversion solves for S-impedance and density, in addition to P-impedance. While many geologic features can express similar P-impedance characteristics, few will share combined P-impedance and S-impedance traits (allowing improved separation and clarity). Often a feasibility study using the wells logs will indicate whether separation of the desired lithotype can be achieved with P-impedance alone or whether S-impedance is also required. This will dictate whether a pre- or post-stack inversion is needed.

Simultaneous Inversion (SI) is a pre-stack method that uses multiple offset or angle seismic sub-stacks and their associated wavelets as input; it generates P-impedance, S-impedance and density as outputs (although the density output resolution is rarely as high as the impedances). This helps improve discrimination between lithology, porosity and fluid effects. For each input partial stack, a unique wavelet is estimated. All models, partial stacks and wavelets are input to a single inversion algorithm —enabling inversion to effectively compensate for offset-dependent phase, bandwidth, tuning and NMO stretch effects.

The inversion algorithm works by first estimating angle-dependent P-wave reflectivities for the input-partial stacks. Next, these are used with the full Zoeppritz equations (or approximations, such as Aki-Richards, for some algorithms) to find band-limited elastic reflectivities. These are in turn merged with their low-frequency counterparts from the model and integrated to elastic properties. This approximate result is then improved in a final inversion for P-impedance, S-impedance and density, subject to various hard and soft constraints. One constraint can control the relation between density and compressional velocity; this is necessary when the range of angles is not great enough to be diagnostic of density.

An important part in the inversion procedure is the estimation of the seismic wavelets. This is accomplished by computing a filter that best shapes the angle-dependent well log reflection coefficients in the region of interest to the corresponding offset stack at the well locations. Reflection coefficients are calculated from P-sonic, S-sonic and density logs using the Zoeppritz equations. The wavelets, with amplitudes representative of each offset stack, are input directly into the inversion algorithm. Since a different wavelet is computed for each offset volume, compensation is automatically done for offset-dependent bandwidth, scaling and tuning effects. A near-stack wavelet can be used as the starting point for estimating the far-angle (or offset) wavelet.

No prior knowledge of the elastic parameters and density beyond the solution space defined by any hard constraints is provided at the well locations. This makes comparison of the filtered well logs and the inversion outputs at these locations a natural quality control. The lowest frequencies from the inversion are replaced with information from the geologic model since they are poorly constrained by the seismic data. When applied in global mode a spatial control term is added to the objective function and large subsets of traces are inverted simultaneously. The simultaneous inversion algorithm takes multiple angle-stacked seismic data sets and generates three elastic parameter volumes as output.

The resulting elastic parameters are real-rock properties that can be directly related to reservoir properties. The more advanced algorithms use the full Knott-Zoeppritz equations and there is full allowance for amplitude and phase variations with offset. This is done by deriving unique wavelets for each input-partial stack. The elastic parameters themselves can be directly constrained during the seismic inversion and rock-physics relationships can be applied, constraining pairs of elastic parameters to each other. Final elastic-parameter models optimally reproduce the input seismic, as this is part of the seismic inversion optimization.

Post-Stack Log Detail Inversion

Geostatistical inversion integrates high resolution well data with low resolution 3-D seismic, and provides a model with high vertical detail near and away from well control. This generates reservoir models with geologically-plausible shapes, and provides a clear quantification of uncertainty to assess risk. Highly-detailed petrophysical models are generated, ready for input to reservoir-flow simulation.

Geostatistics differs from statistics in that it recognises that only certain outcomes are geologically plausible. Geostatistical inversion integrates data from many sources and creates models that have greater resolution than the original seismic, match known geological patterns, and can be used for risk assessment and reduction.

Seismic, well logs and other input data are each represented as a probability density function (PDF), which provides a geostatistical description based on histograms and variograms. Together these define the chances of a particular value at a particular location, and the expected geological scale and composition throughout the modelled area. Unlike conventional inversion and geomodelling algorithms, geostatistical inversion takes a one-step approach, solving for impedance and discrete property types or lithofacies at the same time. Taking this approach speeds the process and improves accuracy.

Individual PDFs are merged using bayesian inference techniques, resulting in a posterior PDF conditioned to the whole data set. The algorithm determines the weighting of each data source, eliminating potential bias. The posterior PDF is then input to a Markov Chain Monte Carlo algorithm to generate realistic models of impedance and lithofacies, which are then used to co-simulate rock properties such as porosity. These processes are typically iterated until a model emerges that matches all information. Even with the best model, some uncertainty remains. Uncertainty can be estimated using random seeds to generate a range of realisations. This is especially useful when dealing with parameters that are sensitive to change; an analysis of this sort enables greater understanding of development risk.

Pre-Stack Log-Detail Inversion

Amplitude versus offset (AVO) (AVA) geostatistical inversion incorporates simultaneous AVO (AVA) inversion into the geostatistical inversion algorithm so high resolution, geostatistics, and AVO may be attained in a single method. The output model (realisations) are consistent with well log information, AVO seismic data, and honour rock property relationships found in the wells. The algorithm also simultaneously produces elastic properties (P-impedance, S-impedance and density) and lithology volumes, instead of sequentially solving for lithology first and then populating the cell with impedance and density values. Because all output models match all input data, uncertainty can be quantitatively assessed to determine the range of reservoir possibilities within the constraining data.

AVA geostatistical inversion software uses leading-edge geostatistical techniques, including Markov Chain Monte Carlo (MCMC) sampling and pluri-Gaussian lithology modelling. It is thus possible to exploit "informational synergies" to retrieve details that deterministic inversion techniques blur out or omit. As a result, geoscientists are more successful in reconstructing both the overall structure and the fine details of the reservoir. The use of multiple-angle-stack seismic volumes in AVA geostatistical inversion enables further evaluation of elastic rock properties and probable lithology or seismic facies and fluid distributions with greater accuracy.

The process begins with a detailed petrophysical analysis and well log calibration. The calibration process replaces unreliable and missing sonic and density measurements with synthesized values from calibrated petrophysical and rock-physics models. Well log information is used in the inversion process to derive wavelets, supply the low frequency component not present in the seismic data, and to verify and analyse the final results. Next, horizon and log data are used to construct the stratigraphic framework for the statistical information to build the models. In this way, the log data is only used for generating statistics within similar rock types within the stratigraphic layers of the earth.

Wavelet analysis is conducted by extracting a filter from each of the seismic volumes using the well elastic (angle or offset) impedance as the desired output. The quality of the inversion result is dependent upon the extracted seismic wavelets. This requires accurate p-sonic, s-sonic and density logs tied to the appropriate events on the seismic data. Wavelets are extracted individually for each well. A final "multi-well" wavelet is then extracted for each volume using the best individual well ties and used as input to the inversion. Histograms and variograms are generated for each stratigraphic layer and lithology, and preliminary simulations are run on small areas. The AVA geostatistical inversion is then run to generate the desired number of realisations, which match all the input data. The results are quality controlled by direct comparison of the inverted rock property volumes against the well logs. Further QC involves review by a multidisciplinary team of all input parameters and the results of the simulation. Analysis of multiple realisations produces mean (P50) property cubes or maps. Most often these are lithology or seismic facies cubes and predicted lithology or facies probabilities, but other outputs are also possible. Selected lithology and facies cubes are also generated for P15 and P85 probabilities (for example). Reservoir 3-D bodies of hydrocarbon-bearing units are captured with their corresponding rock properties, and the uncertainty in reservoir size and properties is quantified.

Offshore Drilling

Offshore drilling refers to a mechanical process where a wellbore is drilled through the seabed. It is typically carried out in order to explore for and subsequently produce hydrocarbons which lie in rock formations beneath the seabed. Most commonly, the term is used to describe drilling activities on the continental shelf, though the term can also be applied to drilling in lakes, inshore waters and inland seas. Offshore drilling presents environmental challenges, both from the produced hydrocarbons and the materials used during the drilling operation. Controversies include the ongoing US offshore drilling debate.

There are many different types of facilities from which offshore drilling operations take place. These include bottom founded drilling rigs (jackup barges and swamp barges), combined drilling and production facilities either bottom founded or floating platforms and deepwater mobile offshore drilling units (MODU) including semi-submersibles and drillships. These are capable of operating in water depths up to 10,000 ft. In shallower waters the mobile units are anchored to the seabed, however in deeper water (>5,000 ft) the semisubmersibles or drillships are maintained at the required drilling location using dynamic positioning.

History

Around 1891, the first submerged oil wells were drilled from platforms built on piles in the fresh waters of the Grand Lake St. Marys (a.k.a. Mercer County Reservoir) in Ohio. The wells were developed by small local companies such as Bryson, Riley Oil, German-American, and Banker's Oil.

Around 1896, the first submerged oil wells in salt water were drilled in the portion of the Summerland field extending under the Santa Barbara Channel in California. The wells were drilled from piers extending from land out into the channel. Other notable early submerged drilling activities occurred on the Canadian side of Lake

Erie in the 1900s and Caddo Lake in Louisiana in the 1910s. Shortly thereafter wells were drilled in tidal zones along the Texas and Louisiana gulf coast. The Goose Creek Oil Field near Baytown, Texas is one such example. In the 1920s drilling activities occurred from concrete platforms in Venezuela's Lake Maracaibo.

One of the oldest subsea wells is the Bibi Eibat well, which came on stream in 1923 in Azerbaijan. The well was located on an artificial island in a shallow portion of the Caspian Sea. In the early 1930s, the Texas Co., later Texaco (now Chevron) developed the first mobile steel barges for drilling in the brackish coastal areas of the Gulf of Mexico.

In 1937, Pure Oil (now Chevron) and its partner Superior Oil (now ExxonMobil) used a fixed platform to develop a field 1 mile offshore of Calcasieu Parish, Louisiana in 14 feet of water.

In 1946, Magnolia Petroleum (now ExxonMobil) drilled at a site 18 miles off the coast, erecting a platform in 18 feet of water off St. Mary Parish, Louisiana.

In early 1947, Superior Oil erected a drilling and production platform in 20 feet of water some 18 miles off Vermilion Parish, La. But it was Kerr-McGee Oil Industries (now Anadarko Petroleum), as operator for partners Phillips Petroleum (ConocoPhillips) and Stanolind Oil & Gas (BP) that completed its historic Ship Shoal Block 32 well in October 1947, months before Superior actually drilled a discovery from their Vermilion platform farther offshore. In any case, that made Kerr-McGee's well the first oil discovery drilled out of sight of land.

When offshore drilling moved into deeper waters of up to 100 feet, fixed platform rigs were built, until demands for drilling equipment was needed in the 100- to 400-foot depth of the Gulf of Mexico, the first jack-up rigs began appearing from specialised offshore drilling contractors such as forerunners of ENSCO International.

The first semi-submersible resulted from an unexpected observation in 1961. Blue Water Drilling Company owned and operated the four-column submersible Blue Water Rig No.1 in the Gulf of Mexico for Shell Oil Company. As the pontoons were not sufficiently buoyant to support the weight of the rig and its consumables, it was towed between locations at a draught mid-way between the top of the pontoons and the underside of the deck. It was noticed that the motions at this draught were very small, and Blue Water Drilling and Shell jointly decided to try operating the rig in the floating mode. The concept of an anchored, stable floating deep-sea platform had been designed and tested back in the 1920's by Edward Robert Armstrong for the purpose of operating aircraft with an invention known as the 'seadrome'. The first purpose-built drilling semi-submersible *Ocean Driller* was launched in 1963. Since then, many semi-submersibles have been purpose-designed for the drilling industry mobile offshore fleet. The first offshore drillship was the *CUSS 1* developed for the Mohole project to drill into the Earth's crust.

As of June, 2010, there were over 620 mobile offshore drilling rigs (Jackups, semisubs, drillships, barges) available for service in the competitive rig fleet.

The world's deepest platform is currently the Perdido in the Gulf of Mexico, floating in 2,438 metres of water. It is operated by Royal Dutch Shell and was built at a cost of $3 billion.

Main Offshore Fields

Notable offshore fields today are found in the North Sea, the Gulf of Mexico, the Campos and Santos Basins off the coasts of Brazil, Newfoundland and Nova Scotia (Atlantic Canada), several fields off West Africa most notably west of Nigeria and Angola, as well as offshore fields in South East Asia and Sakhalin, Russia. Also major offshore oil fields are located in the Persian Gulf such as Safaniya, Manifa, and Marjan which belong to Saudi Arabia and are developed by Saudi Aramco.

Challenges

Offshore oil and gas production is more challenging than land-based installations due to the remote and harsher environment. Much of the innovation in the offshore petroleum sector concerns overcoming these challenges, including the need to provide very large production facilities. Production and drilling facilities may be very large and a large investment, such as the Troll A platform standing on a depth of 300 metres. Another type of offshore platform may float with a mooring system to maintain it on location. While a floating system may be lower cost in deeper waters than a fixed platform, the dynamic nature of the platforms introduces many challenges for the drilling and production facilities.

The ocean can add several hundred metres or more to the fluid column. The addition increases the equivalent circulating density and downhole pressures in drilling wells, as well as the energy needed to lift produced fluids for separation on the platform. The trend today is to conduct more of the production operations subsea, by separating water from oil and re-injecting it rather than pumping it up to a platform, or by flowing to onshore, with no installations visible above the sea. Subsea installations help to exploit resources at progressively deeper waters, locations which have been inaccessible, and overcome challenges posed by sea ice, such as in the Barents Sea.

Offshore manned facilities also present logistics and human resources challenges. An offshore oil platform is a small community in itself with cafeteria, sleeping quarters, management, and other support functions. In the North Sea, staff members are transported by helicopter for a two-week shift. They usually receive higher salary than onshore workers do. Supplies and waste are transported by ship, and the supply deliveries need to be carefully planned because storage space on the platform is limited. Today, much effort goes into relocating as many of the personnel as possible onshore, where management and technical experts are in touch with the platform by video conferencing. An onshore job is also more attractive for the aging workforce in the petroleum industry, at least in the western world. These efforts among others are contained in the established term integrated operations. The increased use of subsea facilities helps achieve the objective of keeping more workers onshore. Subsea facilities are also easier to expand, with new separators or different modules for different oil types, and are not limited by the fixed floor space of an above-water installation.

Effects on the Environment

Offshore oil production involves environmental risks, most notably oil spills from oil tankers or pipelines transporting oil from the platform to onshore facilities, and from leaks and accidents on the platform. Produced water is also generated, which is excess water from oil or gas production and includes varying amounts of oil, or other chemicals used in, or resulting from, oil production. According to the organisation Culture Change, a Gulf of Mexico rig dumps about 90,000 tons of drilling fluid and cuttings over its lifetime, with its wells also contributing with heavy metals.

Oil Well Control

Oil well control is the management of the dangerous effects caused by unexpected high pressures upon surface equipment of oil or gas drilling rigs. Technically, oil well control involves preventing Formation fluid, usually referred to as kick, from entering into the Wellbore during drilling. Formation fluid can enter the wellbore if the pressure exerted by the column of drilling fluid is not great enough to overcome the pressure exerted by the fluids in the formation being drilled. Oil well control also includes monitoring a well for signs of impending influx of formation fluid into the wellbore during drilling, and procedures to stop the well from flowing when it happens by taking proper remedial actions.

Failure to manage and control these pressure effects can cause serious equipment damage and injury/loss of life. Well control situations that are improperly managed cause blowouts, which are the uncontrolled and explosive expulsion of formation fluid from the well, usually resulting in a fire.

Oil well control requires proper understanding of the various pressures that come to play during oil well drilling and also understanding kick concept, when it may occur, and how to prevent it from developing into a blowout.

Importance of Oil Well Control

Oil Well control is one of the most important aspects of drilling operations. Improper handling of kicks in oil well control can result in blowouts with very grave consequences, including the loss of valuable resources. Even though the cost of a blowout (as a result of improper/ no oil well control) can easily reach several millions of US dollars, the monetary loss is not as serious as the other damages that can occur: irreparable damage to the environment, waste of valuable resources, ruined equipment, and most importantly, the safety and lives of personnel on the drilling rig. In other to avert these terrible consequences of blowout, the utmost attention must be given to oil well control. Therefore oil well control procedures should not start when abnormal situation is noticed in the wellbore. Instead, oil well control should start immediately, the moment a new rig position is sited. This will mean from the time the new location is picked, all drilling, completion, workover, snubbing and all other related drilling related operations should be done with proper oil well control in mind.

This will involve widespread training of personnel, the development of strict operational guidelines and the design of drilling programs from the standpoint of maximising the probability of successfully regaining hydrostatic control of a well after a significant influx of formation fluid has taken place.

Fundamental Concepts and Terminology

Pressure is a very important concept in the oil and gas industry. It can be defined as the force exerted per unit area. Its SI unit is newtons per square metre or pascal. Another unit, bar, is also widely used as a measure of pressure, with 1 bar equal to 100 kilopascals. Normally pressure is measured in the petroleum industry in units of pounds force per square inch of area or psi. 1 psi equals 6894.76 pascals.

Hydrostatic Pressure

Hydrostatic pressure(HSP) can be defined as pressure due to a column of fluid that is not moving. That is, a column of fluid that is static or at rest exerts pressure due to local force of gravity on the column of the fluid. The formula for calculating hydrostatic pressure in SI units is:

Hydrostatic pressure = Height (m) × Density (kg/m^3) × Gravity (m/s^2).

All fluids in a wellbore exert hydrostatic pressure, which is a function of density and vertical height of the fluid column. In US oil field units, hydrostatic pressure can be expressed as:

HSP = 0.052 × MV × TVD, where MV (Mud Weight or density) is the drilling fluid density in pounds per gallon ppg, TVD is the true vertical depth in feet and HSP is the hydrostatic pressure in psi. The 0.052 is needed as the conversion factor to psi unit of HSP.

To convert these units to SI units, one can use:

- 1 ppg = H" 119.8264273 kg/m^3
- 1 ft = 0.3048 metres
- 1 psi = 0.0689475729 bar
- 1 bar = 10^5 pascals

Pressure Gradient

Pressure gradient is described as the pressure per unit length. Often in oil well control pressure exerted by fluid is expressed in terms of its pressure gradient. The SI unit is pascal/metres. Hydrostatic pressure gradient can be written as:

Pressure gradient (psi/ft) = HSP/TVD = 0.052 × MV (ppg).

Formation Pressure

Formation pressure is the pressure exerted by the formation fluids which are the liquids and gases contained in the geologic formations encountered while drilling for oil or gas. It can also be said to be the pressure contained within the pores of the formation or reservoir being drilled. Formation pressure is as a result of the hydrostatic pressure of the formation fluids above the depth of interest together with pressure trapped in the formation. Under formation pressure we can have:

- Normally pressured formation
- Abnormal formation pressure
- Subnormal formation pressure

Normally Pressured Formation: Normally pressured formation has a formation pressure that is the same with the hydrostatic pressure of the fluids above it. As the fluids above the formation is usually some form of water, we can define this pressure as the pressure exerted by a column of water from the formation's depth to sea level.

The normal hydrostatic pressure gradient for freshwater is 0.433 pounds per square inch per foot (psi/ft), or 9.792 kilopascals per metre (kPa/m), and 0.465 psi/ft for water with dissolved solids like those in Gulf Coast or 10.516 kPa/m. The density of formation water in saline or marine environments such as the Gulf Coast is about 9.0 ppg or 1078.43 kg/m^3. Since this is the highest for both Gulf Coast water and fresh water, a normally pressured formation can be controlled with a 9.0 ppg mud. Sometimes the weight of the overburden, which is the rocks and fluids above the formation, will tend to compact the formation, resulting in pressure built-up within the formation if the fluids are trapped in place. The formation in this case will retain its normal pressure only if there is a communication with the surface, otherwise an *abnormal formation pressure* will result.

Abnormal Formation Pressure: As discussed above, once the fluids are trapped within the formation and not allow to escape there is a pressure build-up leading to abnormally high formation pressures. This will generally require a mud weight of greater than 9.0 ppg to control. Excess pressure, called overpressure or geopressure, can cause a well to blow out or become uncontrollable during drilling.

Subnormal Formation Pressure: Subnormal formation pressure is a formation pressure that is less than the normal pressure for the given depth. It is common in formations that had undergone production of original hydrocarbon or formation fluid in them.

Overburden Pressure

Overburden pressure is the pressure exerted by the weight of the rocks and contained fluids above the zone of interest. Overburden pressure varies in different regions and formations. It is the force that tends to compact a formation vertically. The density of these usual ranges of rocks is about 18 to 22 ppg. This range of densities will generate an overburden pressure gradient of about 1 psi/ft (22.7 kPa/m). (The 1 psi/ft is not applicable for shallow marine sediments or massive salt.) In offshore however, there is a lighter column of sea water and column of rock do not go all the way to the surface therefore a lower overburden pressure is usually generated at an offshore depth than would be found at the same depth on land.

Mathematically, overburden pressure can be derived as:

$$S = \rho_b \times D$$

where

S = overburden pressure

ρ_b = average formation bulk density

D = vertical thickness of the overlying sediments

The bulk density of the sediment is a function of rock matrix density, porosity within the confines of the pore spaces, and porefluid density. This can be expressed as

$$\rho_b = \varphi\rho_f + (1 - \varphi)\rho_m$$

where

φ = rock porosity

ρ_f = formation fluid density

ρ_m = rock matrix density

Fracture Pressure

Fracture pressure can be defined as pressure required to cause a formation to fail or split. As the name implies, it is the pressure that causes the formation to fracture and the circulating fluid to be lost. Fracture pressure is usually expressed as a gradient, with the common units being psi/ft (kg/m) or ppg (kPa).

To fracture a formation, three things are generally needed, which are:

1. Pump into the formation. This will require a pressure in the wellbore greater than formation pressure
2. The pressure in the wellbore must also exceed the rock matrix strength
3. And finally the wellbore pressure must be greater than one of the three principal stresses in the formation

Pump Pressure (System Pressure Losses)

Pump pressure, which is also referred to as system pressure loss, is the sum total of all the pressure losses from the oil well surface equipment, the drill pipe, the drill collar, the drill bit, and annular friction losses around the drill collar and drill pipe.

It measures the system pressure loss at the start of the circulating system and measures the total friction pressure.

Slow Pump Pressure (SPP)

Slow pump pressure is the circulating pressure (pressure used to pump fluid through the whole active fluid system, including the borehole and all the surface tanks that constitute the primary system during drilling) at a reduced rate. SPP is very important during a well kill operation in which circulation (a process in which drilling fluid is circulated out of the suction pit, down the drill pipe and drill collars, out the bit, up the annulus, and back to the pits while drilling proceeds) is done at a reduced rate to allow better control of circulating pressures and to enable the mud properties (density and viscosity) to be kept at desired values. The slow pump pressure can also be referred to as kill rate pressure, slow circulating pressure, kill speed pressure and so on.

Shut-in Drill Pipe Pressure

Shut-in drill pipe pressure (SIDPP), which is recorded when a well is shut in on a kick, is a measure of the difference between the pressure at the bottom of the hole and the hydrostatic pressure (HSP) in the drillpipe. During a well shut-in, the pressure of the wellbore stabilises and the formation pressure equals the pressure at the bottom of the hole. The drillpipe at this time should be full of known-density fluid, therefore the formation pressure can be easily calculated using the SIDPP. This means that the SIDPP gives a direct of formation pressure during a kick.

Shut-in Casing Pressure (SICP)

The shut-in casing pressure (SICP) is a measure of the difference between the formation pressure and the HSP in the annulus when a kick occurs.

The pressures encountered in the annulus can be estimated using the following mathematical equation:

$$FP = HSP_{mud} + HSP_{influx} + SICP$$

where

FP = formation pressure (psi)

HSP_{mud} = Hydrostatic pressure of the mud in the annulus (psi)

HSP_{influx} = Hydrostatic pressure of the influx (psi)

SICP = shut-in casing pressure (psi).

Bottom-Hole Pressure (BHP)

Bottom-hole pressure (BHP) is the pressure at the bottom of a well. The pressure is usually measured at the bottom of the hole. This pressure may be calculated in a static, fluid-filled wellbore with the equation:

$$BHP = D \times \rho \times C,$$

where

BHP = bottom-hole pressure

D = the vertical depth of the well

ρ = density

C = units conversion factor

(or, in the English system, BHP = D × MWD × 0.052).

The bottom-hole pressure is dependent on the following

- Hydrostatic pressure (HSP)
- Shut-in surface pressure (SIP)
- Friction pressure
- Surge pressure (occurs when transient pressure increases the bottom-hole pressure)
- Swab pressure (occurs when transient pressure reduces the bottom-hole pressure)

Therefore BHP can be said to be the sum of all pressures at the bottom of the wellhole, which equals:

BHP = HSP + SIP + friction + Surge – swab

Basic Calculations in Oil Well Control

There are some basic calculations that need to be carried during oil well control. A few of these essential calculations will be discussed below. Please note that most of the units here are in US oil field units, but these units can be converted to their SI units equivalent by using this Conversion of units link.

Capacity

Here will we look at the capacity of Drill string which is essential in oil well control. The capacity of drillpipe, drill collars or hole is the volume of fluid that can be contained within them.

The capacity formula is as shown below:

$$Capacity = ID^2/1029.4$$

where

Capacity = Volume in barrels per foot(bbl/ft)

ID = Inside diametre in inches

1029.4 = Units conversion factor

Also the total pipe or hole volume is given by :

$$Volume\ in\ barrels(bbls) = Capacity(bbl/ft) \times length(ft)$$

Feet of pipe occupied by a given volume is given by:

Feet of pipe(ft) = Volume of mud(bbls)/ Capacity(bbls/ft)

Capacity calculation is important in oil well control due to the following :

- Volume of the drillpipe and the drill collars must be pumped to get kill weight mud to the bit during kill operation
- It is used to spot pills and plugs at various depths in the wellbore.Annular capacity

This is the volume contained between the inside diametre of the hole and the outside diametre of the pipe. Annular capacity is given by :

$$Annular\ capacity(bbl/ft) = (ID_{hole} - OD_{pipe}{}^{2})/1029.4$$

where

$ID_{hole}{}^{2}$ = Inside diametre of the casing or open hole in inches

$OD_{pipe}{}^{2}$ = Outside diametre of the pipe in inches

Similarly

Annular volume(bbls) = Annular capacity(bbl/ft) × length(ft)

and

Feet occupied by volume of mud in annulus = Volume of mud(bbls)/ Annular Capacity(bbls/ft).

Fluid Level Drop

Fluid level drop is the distance the mud level will drop when a dry string (a bit that is not plugged) is being pulled from the wellbore and it is given by :

Fluid level drop = Bbl disp/(CSG cap - Pipe disp)

or

Fluid level drop = Bbl disp/(Ann cap + Pipe disp)

and the resulting loss of HSP is given by:

Lost HSP = 0.052 × MV × Fluid drop

where

Fluid drop = Distance the fluid falls (ft)

Bbl disp = Displacement of the pulled pipe (bbl)

CSG cap = Casing capacity(bbl/ft)

Pipe disp = pipe displacement(bbl/ft)

Ann cap = Annular capacity between casing and Pipe(bbl/ft)

Pipe cap = Lost hydrostatic pressure(psi)

Lost HSP = Lost hydrostatic pressure(psi)

MW = Mud Weight(ppg)

When pulling a wet string(the bit is plugged) and the fluid from the drillpipe is not returned back to the hole. The Fluid drop is then changed to the following:

$$Fluid\ level\ drop = Bbl\ disp / Ann\ cap$$

Kill Weight Fluid

Kill weight fluid which can also be called *Kill weight Mud* is the density of the mud required to balance formation pressure during kill operation. The Kill Weight Mud can be calculated by:

$$KVM = SIDPP/(0.052 \times TVD) + OWM$$

where

KVM = Kill Weight Mud(ppg)

SIDPP = Shut-In drillpipe pressure(psi)

TVD = True Vertical depth(ft)

OVM = Original weight mud(ppg)

But when the formation pressure can be determined from data sources such as bottom hole pressure, then *KWM* can be calculated as follows:

$$KVM = FP/0.052 \times TVD$$

FP = Formation pressure.

Kicks

Kick is the entry of formation fluid into the wellbore during drilling operations. It occurs because the pressure exerted by the column of drilling fluid is not great enough to overcome the pressure exerted by the fluids in the formation drilled. The whole essence of oil well control is to prevent kick from occurring and if it happens to prevent it from developing into blowout. An uncontrolled kick must times are as a result of proper equipment not deployed, poor practices and lack of training of the rig crews. Loss of oil well control may lead into blowout, which represent one of the most severe threat associated with exploration of petroleum resources involving the risk of lives and environmental and economic consequences.

Causes of Kicks

A kick will occur when the bottom hole pressure(BHP) of a well falls below the formation pressure and the formation fluid flows into the wellbore. There are usually causes for kicks some of which are:

- Failure to keep the Hole full during a Trip
- Swabbing while tripping
- Lost circulation
- Insufficient density of fluid
- Abnormal pressure
- Drilling Into an Adjacent Well
- Lost control during drill stem test.

Failure to Keep the Hole Full During a Trip

Tripping is the complete operation of removing the drillstring from the wellbore and running it back in the hole. This operation is typically undertaken when the bit (which is the tool used to crush or cut rock during drilling) becomes dull or broken, and no longer drills the rock efficiently. Tripping out of the hole means a volume of steel (of drillstring) is been removed from the well. This displacement of the drill string(the steel) will leave out a volume of space that must be replaced with equal volume of mud. If the replacement is not done, the fluid level in the wellbore will drop resulting in a loss of hydrostatic pressure (HSP) and bottom hole pressure (BHP). If this bottom hole pressure reduction goes below the formation pressure, a kick will definitely occur.

Swabbing While Tripping

Swabbing occurs when bottom hole pressure is reduced due to the effects of pulling the drill string or upward movement of pipe in a well. During tripping out of the hole, the space formerly by the drillpipe, drill collar, or tubing which are being removed must be replaced by something usually mud. If the rate of tripping out is greater than the rate the mud is being used to fill the void space created by the removal of the drill string, then swab will occur. If the reduction in bottom hole pressure caused by swabbing is below formation pressure, then a kick will occur.

Lost Circulation

Lost circulation usually occurs when the hydrostatic pressure fractures an open formation. When this occurs there is loss in circulation and the height of the fluid column decreases leading to lower HSP in the wellbore. A kick can occur if steps are not taken to keep the hole full. Lost circulation can be caused by:

- Excessive mud weights
- Excessive annular friction loss
- Excessive surge pressure during trips or spudding the bit
- Excessive shut-in pressures

Insufficient Density of Fluid

If the density of the drilling fluid or mud in the well bore is not sufficient to keep the formation pressure in check then a kick can occur. Insufficient density of the drilling fluid can be as a result of the following :

- attempting to drill underbalanced
- excessive dilution of the mud
- heavy rains in the pits
- barite settling in the pits
- spotting low density pills in the well

Abnormal Pressure

Another cause of kicks is drilling accidentally into abnormally pressured permeable zones. The increased formation pressure may be greater than the bottom hole pressure resulting in a kick.

Drilling into an Adjacent Well

Drilling into an adjacent well is a potential problem, particularly offshore where a large number of directional wells are drilled from the same platform. If the drilling well penetrates the production string of a previously completed well, the formation fluid from the completed well will enter the wellbore of the drilling well, causing a kick. If this occurs at a shallow depth, it is an extremely dangerous situation and could easily result in an uncontrolled blowout.

Lost Control During Drill Stem Test

A drill stem test is performed by setting a packer above the formation to be tested, and allowing the formation to flow. During the course of the test, the bore hole or casing below the packer, and at least a portion of the drill pipe or tubing, is filled with formation fluid. At the conclusion of the test, this fluid must be removed by proper well control techniques to return the well to a safe condition. Failure to follow the correct procedures to kill the well could lead to a blowout.

Kick Warning Signs

In oil well control a kick should be able to be detected promptly and proper kick prevention operation taken immediately to avoid blowout. There are various signs that tell-tale signs a kick will give, knowing these signs will allows oil well control to be carried out promptly to avert blowout.

Sudden Increase in Drilling Rate

A sudden increase in penetration rate (drilling break) is usually caused by a change in the type of formation being drilled: however, it may also signal an increase in formation pore pressure which may indicate a possible kick.

Increase in Annulus Flow Rate

If the rate at which the pumps are running is held constant, the flow from the annulus should be constant. If the annulus flow increases without a corresponding change in pumping rate, the additional flow is caused by formation fluid(s) feeding into the wellbore or gas expansion. This will indicate an impending kick.

Gain in Pit Volume

If there is an unexplained increase in volume of surface mud pit (a large tank that holds drilling fluid on the rig) it could signifies an impending kick. This is because as the formation fluid feeds into the wellbore, it causes more drilling fluid to flow from the annulus than is pumped down the drill string, thus the volume of fluid in the pit(s) increases.

Change in Pump Speed or Pressure

A decrease in pump pressure or increase in pump speed is as a result of a decrease in hydrostatic pressure of the annulus as the formation fluids enters the wellbore. As the lighter formation fluid flows into the wellbore, the hydrostatic pressure exerted by the annular column of fluid decreases, and the drilling fluid in the drill pipe tends to U-tube into the annulus. When this occurs, the pump pressure will drop and the pump speed will increase. The lower pump pressure and increase in pump speed symptoms are also indicative of a hole in the drill string, commonly referred to as a washout. Until a confirmation can be made whether a washout or a well kick has occurred, a kick should be assumed.

Improper Fill on Trips

Improper fill on trip occurs when volume of drilling fluid to Keep the hole full on a Trip (complete operation of removing the drillstring from the wellbore and running it back in the hole) is less than that calculated or less than Trip Book Record. This condition is usually caused by formation fluid entering the wellbore due to the swabbing action of the drill string and if action is not taken soon, the well will enter a kick state.

Categories of Oil Well Control

There are basically there types of oil well control which are:

- Primary oil well control
- Secondary oil well control
- Tertiary oil well Control

Primary Oil Well Control

Primary oil well control is the process which maintains a hydrostatic pressure in the wellbore greater than the pressure of the fluids in the formation being drilled, but less than formation fracture pressure. It uses the mud weight to provide sufficient pressure to prevent an influx of formation fluid into the wellbore If hydrostatic pressure is less than formation pressure then formation fluids will enter the wellbore.

If the hydrostatic pressure of the fluid in the wellbore exceeds the fracture pressure of the formation then the fluid in the well could be lost. In an extreme case of lost circulation the formation pressure may exceed hydrostatic pressure allowing formation fluids to enter into the well.

Secondary Oil Well Control

Secondary oil well control is done after the Primary oil well control has failed to prevent formation fluids entering the wellbore. This process is stopped using a "blow out preventer", BOP to prevent the escape of wellbore fluids from the well.

Tertiary Oil well Control

Tertiary oil well control describes the third line of defence. Where the formation cannot be controlled by primary or secondary well control (hydrostatic and equipment). This happens in Underground blowout situations. These following examples are tertiary well control:

- Drill a relief well to hit an adjacent well that is flowing and kill the well with heavy mud
- Rapid pumping of heavy mud to control the well with equivalent circulating density

- Pump barite or heavy weighting agents to plug the wellbore in order to stop flowing
- Pump cement to plug the wellbore.

Shut-in Procedures

Shut-in Procedures is one of the oil well control measure to curtail kicks and prevent blowout from occurring. Shut in procedures are specific procedures for closing a well in case of well control situation. When any positive indication of a kick is observed such as a sudden increase in flow or an increase in pit level, then the well should be shut in immediately. If a well shut-in is not done promptly, a blowout is likely to happen.

Shut-in procedures are usually developed and practiced for every rig activity such as, drilling, tripping, logging, running tubular, performing a drill stem test and so on and the main reason to have the specific shut in procedure is to minimise kick volume entering into a wellbore when well control situation occurs. It should however be noted that shut in procedure is a company specific procedure and the policy of a company will dictate a well should be shut-in.

They are generally two type of Shut-in procedures which are:

- Soft Shut-In
- Hard Shut-In

Among the two methods, the hard shut-in is the fastest method to shut in the well; therefore, it will minimise volume of kick allowed into the wellbore.

Well Kill Procedures

Well kill procedure is an oil well control method. Once the well has been shut-in on a kick, proper kill procedures must be done immediately. The general idea in well kill procedure is to circulate out any formation fluid already in the wellbore during kick and circulate a satisfactory weight of kill mud called Kill Weight Mud (KWM) into the well without allowing further fluid into the hole. If this can be done, then once the kill mud has been fully circulated around the well, it is possible to open up the well and restart normal operations. Generally, a kill mud(KWM) which just provides hydrostatic balance for formation pressure is circulated. This allows approximately constant bottom hole pressure which is slightly greater than formation pressure to be maintained as the kill circulation proceeds because of the additional small circulating friction pressure loss. After circulation, the well is opened up again.

The major well kill procedures used in oil well control are listed below:

- Wait and Weight
- Driller method
- Circulate and Weight
- Concurrent Method
- Reverse Circulation
- Dynamic Kill procedure
- Bullheading
- Volumetric Method
- Lubricate and Bleed.

Oil Well Control Incidents - Root Causes

There will always be oil well control problems as long as there are drilling operations anywhere in the world. Most of these well control problems are as a result of some errors and can be eliminated, even though some are actually unavoidable. Since we know the consequences of failed well control are severe, effort should be made to prevent some human errors which are the root causes of these incidents. These causes include:

- Lack of knowledge and skills of rig personnel
- Improper work practices
- Lack of understanding of oil well control training
- Lack of application of policies and standards
- Inadequate Risk Management.

Organisations for Building Oil Well Control Culture

Good oil well control culture requires personnels involve in oil well control to develop a core value for it by doing the proper thing at the proper time. A good well control culture will definitely minimise well control incidents. Building well control culture would involve developing competent personnel that are able to recognise well control problems and know what to do to mitigate against them. This is usually done through quality assured programs and training. These programs are done by organisations such as International Association of Drilling Contractors (IADC) and International Well Control Forum (IWCF).

IADC operates the Well-Control accreditation Program (WellCAP) which is a training aimed at providing the necessary knowledge and practical skills critical to successful well control and to develop competent rig personnel. This training starts with floor-hand level and continued to the most-experienced drilling personnel.

IWCF is an NGO whose main aim is to develop and administer well control certification programmes for personnel employed in oil well Drilling, Workover and Well Intervention operations.

Oil Well Fire

Oil well fires is the common term for oil or gas wells that have caught on fire and burn. Oil well fires can be the result of human actions, such as accidents or arson, or natural events, such as lightning. They can exist on a small scale, such as an oil field spill catching fire, or on a huge scale, as in geyser-like jets of flames from ignited high pressure wells.

Extinguishing the Fires

***Figure:** Kuwaiti firefighters fight to secure a burning oil well in the Iraqi Rumaila oilfields in 2003.*

Oil well fires are more difficult to extinguish than regular fires due to the enormous fuel supply for the fire. In fighting a fire at a wellhead, typically high explosives, such as dynamite, are used to consume all the local atmospheric oxygen and *snuff* the flame out. Doing so removes the oxygen necessary for the fire to burn, but the fire's fuel, whether it be natural gas or oil, is still present and oil can shower down upon the working crew.

After snuffing, the wellhead must be capped to stop the flow of oil. During this time, the fuel and oxygen required to create another inferno is present in copious amounts. At this perilous stage, one small spark (perhaps from a steel or iron tool striking a stone) or other heat source might re-ignite the oil. To prevent re-ignition, brass or bronze tools, which do not strike sparks, or paraffin coated tools are used during the capping process. Meticulous care is used to avoid heat and sparks, or any other ignition source. The explosive re-ignition of a wellhead may take the form of an extremely powerful explosion, possibly even worse than the original blowout.

Due to recent advances in technology as well as environmental concerns, many wells today are capped while they burn. High-powered water sprays and Purple K dry chemical (a potassium bicarbonate mixture) are used to extinguish the wells.

There are several techniques used to put out oil well fires, which vary by resources available and the characteristics of the fire itself.

In essence the trade was started by Myron M. Kinley, who dominated the field in the early years. His lieutenant, Red Adair, went on to become the most famous of oil well firefighters.

Techniques include:

- Dousing with copious amounts of water
- Raising the plume- Inserting one metal casing 30 to 40 feet high over the well head (thus raising the flame above the ground). Liquid nitrogen or water is then forced in at the bottom to reduce the oxygen supply and put out the fire.
- Drill relief wells into the producing zone to redirect some of the oil and make the fire smaller. (However, most relief wells are used to pump heavy mud and cement deep into the wild well.) The first relief wells were drilled in Texas in the mid 1930s.
- Using a gas turbine to blast a fine mist at the fire. Water is injected to the compressor section of the turbine in large quantities. This does not harm the turbine. This technique is also used for cleaning turbines.
- Using dynamite to 'blow out' the fire by blasting fuel and oxygen from the flame and consuming oxygen in the combustion. This was one of the earliest effective methods and is still widely used. The first use was by Myron Kinley's father in California in 1913
- Dry Chemical (mainly Purple K) can be used on small well fires such as those in refineries.

- In the 1930s mechanical jaws were developed for smaller well fires which clamp off the pipe below the fire. Rarely used. The design became the bases for the safety device used on off shore wells.

Special vehicles called "Athey wagons" as well as the typical bulldozer protected by corrugated steel sheeting are normally used in the process.

Effects

Oil well fires can cause the loss of millions of barrels of crude oil per day. Combined with the ecological problems caused by the large amounts of smoke and unburnt petroleum falling back to earth, oil well fires such as those seen in Kuwait can cause enormous economic losses. Smoke from burnt crude oil contains many chemicals, including sulfur dioxide, carbon monoxide, soot, benzopyrene, Poly aromatic hydrocarbons, and dioxins. Exposure to oil well fires is commonly cited as a cause of the Gulf War Syndrome, however, studies have indicated that the firemen who capped the wells did not report any of the symptoms suffered by the soldiers.

Driller (Oil)

The driller is a team leader in charge during the process of well drilling. The term is commonly used in the context of an oil well drilling rig.

The driller is in charge of his crew, and running the rig itself. Most of the time, his job is simply to monitor the rig's activity, while the automatic driller runs the breaks and drills the hole. The driller is responsible for interpreting the signals the well gives regarding gas and fluids with high pressure. In an emergency he is also responsible for taking the correct counter measures to stop an uncontrolled well control situation from emerging. The driller will watch for gas levels coming out of the hole, how much drilling mud is going in and other information. While tripping, the driller will run the floor and work the rig. In the context of an offshore oil platform, the driller will be the one in charge of real time decisions. According to and, the hierarchy on an oil platform correspond to the timescale of which the sections operate. The automatic drilling equipment work on the timescale of seconds and both reports to and gets its orders from the driller, who operates on a timescale of several seconds to hours. He reports and gets his or her orders from those planning the current drilling operation, on a timescale of days and weeks. The chain extends to those who are in charge of managing the whole oil field, on a timescale of decades.

Chapter 2

Oil Well Drilling

With all the focus on oil in the Middle East and elsewhere overseas, it can be easy to overlook that the idea of drilling for oil was an American one. The first oil wells were drilled over a century before OPEC came into being, and long before anyone worried about dwindling fossil fuels. They've come a long ways since then, but modern platforms started humbly.

Petroleum oil was discovered, but wasn't used much, by the late 1850s. People used whale oil for lamps and other oil-based uses, and it wasn't until a find in Titusville, Pennsylvania, of burnable petroleum that those ideas began to change. In 1858, "Colonel" Edwin Drake was sent by the Seneca Oil Company to investigate the find in Pennsylvania. In Titusville he found that the reports about the oil seeping from the ground were true, and he began devising a way to extract it for commercial use.

Drilling for Oil

Drake had knowledge of the drilling techniques used for salt mining, and thought something similar could be adapted for oil. Early efforts met with difficulties, but eventually Drake came up with the method of using pipes to surround the drilling tools, preventing the holes from collapsing.

The First "Gusher"

Drake struck oil, though it did not erupt in the violent gushers most people think of. It seeped to the surface, and was eventually brought up with a pump and poured into buckets. Yet despite the lack of drama, Drake's method allowed for the collection of oil from far below the surface.

Start of an Industry

There are reports of similar attempts at drilling for oil elsewhere in the world, but it was Drake's efforts in Pennsylvania that eventually led to the modern oil industry as we know it today.

Famous Ties

The events surrounding the first well were made into a 1954 movie starring Vincent Price. "Born in Freedom: The Story of Colonel Drake" is shown daily in the Drake Well Museum in Titusville, Pennsylvania

Florida's First Oil Well

Among its records for dry holes, Florida's first – but certainly not last – unsuccessful attempt to find commercially viable oil reserves began in 1901, not far from the Gulf Coast panhandle town of Pensacola. Two test wells were drilled, the first to 1,620 feet and the second a hundred feet deeper. Both were abandoned. Whether that wildcatter was following science or intuition, contemporary accounts of his efforts reveal only a small historical footnote: "Florida's first dry holes." Twenty years later, as America's oil demand continued to soar, oil still had not been found in Florida. The state's panhandle still looked promising – despite a growing list of failed drilling ventures.

Indian legends and a wildcat stock promoter's claim of oil inspired yet another attempt near today's Falling Waters Park, about 100 miles east of Pensacola. A tall, wooden derrick and steam-driven rig were used to drill. At a depth 3,900 feet, a brief showing of natural gas excited area residents with a false report of a possible gusher. Undeterred, the oilmen continued to drill to a depth of 4,912 feet before finally giving up.

No oil of commercial quantity was found and the well was capped in 1921. Another dry hole.

Barron Collier

Around this time, one of Florida's most revered visionaries and entrepreneurs, Barron G. Collier, was busily purchasing land in the sparsely populated southwest part of the state. Between 1921 and 1923, he acquired about 1.3 million acres that would eventually become Collier and Hendry counties, including what is now the "Big Cypress Preserve."

Collier had made his fortune in streetcar advertising sales, beginning in his native Memphis and spreading from New York to San Francisco as Consolidated Street Railway Advertising Company. With his capital and a vision of Florida alien to most, one of his first challenges was construction of the Tamiami Trail (U.S. 41). This road, extending for 368 miles from Tampa southward along the Gulf coast to Naples, then eastward to Miami, was built through some of the most difficult terrain in the United States – mostly dense swamps and wilderness infested with snakes and alligators.

Barron Collier's advocacy and personal financial backing was key to successful completion of the Tamiami Trail in April 1928. Ever the savvy businessman, Collier negotiated his first oil lease in the county with Gulf Oil Company in the mid 1930s, despite Florida's still unbroken string of dry holes.

Gulf Oil brought in 50 men to conduct seismic testing, using the first big-wheeled "swamp buggy" vehicles of their type in the county. Gulf established headquarters in Everglades City, then the county seat, and began the search. For 10 years, Gulf searched. The company drilled a number of wells, some to a depth of about 6,000 feet, but ultimately, seismic tests convinced them that full scale drilling was not warranted and in 1938, Gulf Oil pulled out.

$50,000 Bounty

By 1939, almost 80 dry holes had been drilled, the deepest to a depth of 6,180 feet. About this time, Florida legislators, desperate for their state to become an oil producer, offered a $50,000 bounty for the first discovery.

Hoping to find success at greater depth, Peninsular Oil and Refining Company drilled in Southwest Florida's Monroe County to 10,006 feet, but still found no oil.

Collier's confidence nevertheless remained unshakeable. His son relates of the time, "I said to dad, 'You know, perhaps we have to face the fact that maybe there is no oil in Collier County.' Well, he was just absolutely furious. He shook his finger under my nose and said, 'Just don't let anybody tell you that there isn't any oil in Collier County.' And when I looked at him, he smiled and said, 'I can smell it.'"

Following their Monroe county disappointment, Peninsular executed a lease assignment to Humble Oil and Refining Co. and Humble began searching near the Sunniland watering stop on the Atlantic Coast Line Railroad. Barron Collier remained confident that oil would be found in southwest Florida, but when he died in 1939, oil in Florida remained an unrealised dream.

At Sunniland, the search continued, with the drilling done by the Hoffland Brothers of Tulsa, Oklahoma, one of the foremost drilling companies in the country.

Then, on the 26th of September 1943, after expending about $1 million and reaching a depth of 11,626, Humble Oil Co. brought in Sunniland No. 1, Florida's first producing oil well. The well's site is about 12-miles south of Immokalee, by present day Big Cypress Preserve and a 30-minute drive from the resort city of Naples.

Initial daily production was 140 barrels of oil and 425 gallons of salt water, which eventually settled down to 20 barrels per day. This was no gusher, but it proved the tenacious Barron Collier's wildcatter intuition to have been right on target.

Predictably, the Humble Oil Co. find sparked a flurry of lease purchases and wildcat wells. By 1954, the Sunniland field was producing 500,000 barrels per year from eleven wells at average depths of 11,575 feet. Sunniland remained Florida's top producer until 1964, when Sun Oil Co., after spending $10 million on 34 dry holes, discovered the Felda field in nearby Hendry County. Pleased with its discovery of the Sunniland oilfield, Humble Oil accepted the $50,000 prize offered by the state, added $10,000 – and donated the $60,000 equally between the University of Florida and the Florida State College for women.

Humble would later become the Exxon Corp., now ExxonMobil.

Offshore Oil History

American Oil & Gas Historical Society: America's offshore oil industry began in the Pacific Ocean more than 100 years ago. As recently as 1947 no company had ever risked drilling beyond the sight of land.

In 1896, as enterprising businessmen pursued California's prolific Summerland oilfield all the way to the beach, the lure of offshore production enticed Henry L. Williams and his associates to build a pier 300 feet out into the Pacific — and mount a standard cable-tool rig on it.

By 1897 this first offshore well was producing oil and 22 companies soon joined in the boom, constructing 14 more piers and over 400 wells within the next five years. The Summerland offshore field produced for 25 years — fuelling the growth of California's economy.

In 1911, Gulf Refining Co. abandoned the use of piers. It drilled Ferry Lake No. 1 on Caddo Lake, La., using a fleet of tugboats, barges, and floating pile drivers.

When the well came in at 450 barrels per day, Gulf constructed platforms every 600 feet on each 10-acre lakebed site.

It was the first time drilling was done over water without a pier connection to shore.

In 1938, Pure Oil Co. and Superior Oil Co. built a freestanding drilling platform in the Gulf of Mexico, despite logistics, engineering, and communications challenges. They hired a Houston engineering and construction company, Brown & Root Marine Operators, Inc., to build a 320-foot by 180-foot freestanding wooden deck in 14-feet of water about a mile offshore. The chosen drilling site was near Creole, La. Using onshore building criteria and intuition, the Creole platform was designed to withstand winds of 150 mph and constructed 15-feet above the water. Three hundred treated yellow pine pilings were driven 14 feet into the sandy bottom.

The Superior-Pure State No. 1 well was successful – but was wiped off its pilings by a hurricane in 1940. Although the pilings were damaged, the platform was quickly rebuilt and put back into production in the four million barrel field.

Searching for Gulf of Mexico Salt Domes

> *"It may be tentatively assumed that the Gulf of Mexico is a potential source of salt-dome oil... Whether or not it will ever be economically feasible to explore these waters for the domes that must exist is a question for the future to answer."* *– Geologist Orval Lester Brace in 1941*

Kerr-McGee answered the salt dome question with the Kermac No. 16 offshore rig in 1947.

Not much equipment specifically designed for offshore drilling existed and exploration remained an extraordinarily speculative and risky business venture. An offshore dry hole could easily swallow the huge capital costs sunk into construction of a large, permanent rig platforms.

Nevertheless, Dean McGee of Kerr-McGee Oil Industries Inc. partnered with Phillips Petroleum and Stanolind Oil & Gas Co. to secure leases for exploratory wells in the Gulf of Mexico. They hired Brown & Root to build a freestanding platform 10 miles out to sea.

> *"We decided to explore the areas where the really potential prolific production might be – salt domes – the good ones on land were gone, but we could move out in the shallow water and, in effect, get into a virgin area where we could find the real class-one type salt dome prospect," McGee said.*

Vessels were needed to provide supplies, equipment, and crew quarters for the drilling site, 43 miles southwest of Morgan City, La. The gradually sloping Gulf of Mexico reached only about 18-feet deep at the drilling site.

A second platform would be built about eight miles from the first at Ship Shoal Block 28. Sixteen 24-inch pilings were sunk 104 feet into the ocean floor to secure a 2,700 square foot wooden deck.

The Kermac No. 16 well stood in almost 20 feet of water, ten miles at sea. It was spudded on Sept. 10, 1947. The biggest hurricane of the season arrived a week later – with winds of 140 mph. Kerr-McGee had $450,000 invested in the project. Although both platforms were evacuated during the hurricane, damage was minimal and drilling promptly resumed. On Nov.14, the Kermac No. 16 well came in at 40 barrels per hour. The *Oil & Gas Journal* proclaimed, "Spectacular Gulf of Mexico Discovery...Possible 100-Million Barrel Field – 10 Miles at Sea." Kermac No.16 would produce 1.4 million barrels of oil and 307 million cubic feet of natural gas by 1984.

Modern offshore energy industry benefits come from the hard lessons learned from 60 years of open water experience. Compared to the limits of just a few years ago, today's achievements will no doubt pale in comparison to what the future of offshore exploration will bring.

Editor's Note — By the end of 1949, 11 oil and natural gas fields were found in the Gulf of Mexico with 44 exploratory wells, according to the National Ocean Industries Association, which notes that the industry continued to through the 1950s. Revenue generated from the production of oil became the second-largest revenue generator for the country, after income taxes. NOIA also notes:

"As the industry entered the last decade of the 20th century, advancing technology ensued. New depth records for drilling reached 7,625 feet in the Gulf of Mexico, and Shell Oil's platform 'Troll,' which stands in the North Sea in 1,000 feet of water, 1,500 feet high, became one of two man-made objects visible with the naked eye from the surface of the moon. The other is the Great Wall of China."

Onshore salt domes were recorded as early as 1890 by the Geological Survey of Texas. But early drillers were more interested in the many unexploited natural oil seeps – until the famous 1901 gusher at Spindletop in Beaumont. Seismic instruments have long since revealed how ancient salt moved up through the earth, bending or breaking rock formations – sometimes leaving oil trapped.

The first use of helicopters in transportation to offshore platforms was carried out at the request of Kerr-McGee and Humble Oil & Refining Co. Bell Helicopters had developed the Model 47 and recognised the opportunity by forming Petroleum Bell Helicopters Co.

A flat area on an LST (from WW Two's landing ship, tanks) anchored next to Humble Rig 28 served as landing pad for one of the first helicopters to be flown offshore. At left is a Bell Helicopter advertisment from 1954. – *From the Ocean Star Offshore Drilling Rig and Museum.*

Moveable rigs drill many exploratory offshore wells. Sometimes it is more economical to build a permanent platform from which well completion, extraction and production can occur. These large, permanent platforms are extremely expensive; they generally require large expected hydrocarbon deposits to be economical to construct.

This depiction of offshore drilling and completion platforms gives an idea of just how massive modern rigs can be. Because of their size, most permanent offshore rigs are constructed in pieces near land. As components of the rig are completed, they are taken out to the drilling location. Sometimes construction or assembly can even take place as the rig is being transported to its intended destination.

The History of the Oil Industry

Oil Through the Ages

347 A.D. Oil wells are drilled in China up to 800 feet deep using bits attached to bamboo poles. 1264 Mining of seep oil in medieval Persia witnessed by Marco Polo on his travels through Baku. 1500s

Seep oil collected in the Carpathian Mountains of Poland is used to light street lamps. 1594 Oil wells are hand dug at Baku, Persia up to 35 metres (115 feet) deep. 1735 Oil sands are mined and the oil extracted at Pechelbronn field in Alsace, France. 1802 A 58-ft well is drilled using a spring pole in the Kanawha Valley of West Virginia by the brothers David and Joseph Ruffner to produce brine. The well takes 18 months to drill. 1815 Oil is produced in United States as an undesirable by-product from brine wells in Pennsylvania. 1848 First modern oil well is drilled in Asia, on the Aspheron Peninsula north-east of Baku, by Russian engineer F.N. Semyenov. 1849 Distillation of kerosene from oil by Canadian geologist Dr. Abraham Gesner. Kerosene eventually replaces whale oil as the illuminant of choice and creates a new market for crude oil. 1850 Oil from hand-dug pits in California at Los Angeles is distilled to produce lamp oil by General Andreas Pico. 1854 First oil wells in Europe are drilled 30- to 50-metres deep at Bóbrka, Poland by Ignacy Lukasiewicz. 1854 Natural Gas from a water well in Stockton, California is used to light the Stockton courthouse. 1857 Michael Dietz invents a kerosene lamp that forces whale oil lamps off the market. 1858 First oil well in North America is drilled in Ontario, Canada. 1859 First oil well in United States is drilled 69 feet deep at Titusville, Pennsylvania by Colonel Edwin Drake.

California Comes of Age

1861 First oil well in California is drilled manually in Humboldt County. 1866 Oil is collected from tunnels dug at Sulphur Mountain in Ventura County by the brothers of railroad baron Leland Stanford, the same year that these techniques are applied to the Pechelbronn oil mine in France. 1866 First steam-powered rig in California drills an oil well at Ojai, not far from the Sulphur Mountain seeps. 1875 First commercial oil field in California is discovered at Pico Canyon in Los Angeles County. 1878 Electric light bulb invented by Thomas Edison eliminates demand for kerosene, and the oil industry enters a recession. 1885 Gas wells are drilled in Stockton, California for fuel and lighting. 1885 Oil burners on steam engines in the California oil fields, and later on steam locomotives, create new crude oil markets. 1886 Gasoline-powered automobiles introduced in Europe by Karl Benz and Wilhelm Daimler create additional markets for California oil. Prior to the automobile, gasoline was a cheap solvent produced as a byproduct of kerosene distillation. 1888 A steel-hulled tanker sails from Ventura to San Francisco, eleven years after the 1877 sailing of a Russian tanker across the Caspian sea at Baku. 1899 Discovery of Kern River oil field propels Kern County to top oil-producing region in state.

1860s to 1890s - Tar Pits and Tunnels:

- 1864 - Tar mined from open pits at Asphalto (McKittrick) on west side of San Joaquin Valley.
- 1866 - First refinery in Kern County built near McKittrick tar pits to process kerosene and asphalt.
- 1878 - First wooden derrick in Kern County constructed at Reward to drill for flux oil to mix with asphalt.
- 1887 - "Wild Goose" well at Oil City, Coalinga comes in at 10 bbls/day, demonstrating potential of north part of basin.
- 1889 - Oil wells drilled at Old Sunset (Maricopa) with a steam-powered rig mark discovery of Midway-Sunset field.
- 1893 - Railroad reaches McKittrick, where tunnels and shafts are dug to mine asphalt.
- 1894 - Old Sunset (Maricopa) part of Midway-Sunset has 16 wells producing 30 barrels of oil per day.

1890s to 1920s - Gushers and Cable Tools:

- 1896 - Shamrock Gusher blows in at McKittrick and hastens end of tar mining operations.
- 1899 - Hand-dug oil well discovers Kern River field and starts an oil boom in Kern County.
- 1902 - Arrival of railroad makes development of Midway-Sunset field economically feasible.
- 1902 - First rotary rig in Califonia reportedly drills a well at Coalinga field, but the hole is so crooked that a cable tool is used to redrill the well.
- 1903 - Kern River and Midway-Sunset production makes California the top oil producing state.
- 1904 - 17.2 million bbls of oil produced at Kern River exceeds annual production from Texas.
- 1908 - Rotary drilling rigs and crews arrive in California from Louisiana and successfully drill wells at Midway-Sunset field and erase the embaressment of the Coalinga experiment six years earlier.
- 1909 - Midway Gusher blows out near Fellows and focuses attention on Midway-Sunset field.
- 1910 - Lakeview Gusher blows in near Taft and becomes America's greatest oil gusher.

- 1919 - Hay No. 7 catches fire at Elk Hills and becomes America's greatest gas gusher.
- 1929 - Blowout prevention equipment becomes mandatory on oil and gas wells drilled in California.

1930s to 1950s - Well Logs, Seismic, and Rotary Drilling:

- 1929 - First well logs in California run by Shell in a well near Bakersfield (Kern County).
- 1930 - Deepest well in the world is Standard Mascot #1, rotary drilled to 9,629 feet at Midway-Sunset.
- 1936 - First seismic exploration in California discovers Ten Section field near Bakersfield. Seismic discovery of the productive Paloma and Coles Levee anticlines soon follows
- 1943 - Deepest well in the world is Standard 20-13, drilled to 16,246 feet at South Coles Levee.
- 1953 - Deepest well in the world is Richfield 67-29 drilled to 17,895 feet at North Coles Levee.

1960s to Today - Steam, Horizontal Wells, and Computers:

- 1961 - First steam recovery projects in Kern County start up at Kern River and Coalinga fields after a successful pilot by Shell at Yorba Linda field in Los Angeles.
- 1973 - Tule Elk and Yowlumne fields become the last 100-million barrel fields discovered in Kern County.
- 1980 - First horizontal well in Kern County is Texaco Gerard #6 in fractured schist at Edison field.
- 1980s - Cogeneration hastens the spread of steam recovery projects, which dramatically ramp up oil production.
- 1985 - Kern County reaches an all-time production high of 256 million barrels of oil/year. At the same time, California reaches an all-time production high of 424 million barrels of oil/year.
- 1990s - 3D-seismic data and 3D-computer modelling of reservoirs bring new life to old fields.
- 1997 - Deepest horizontal well in Kern County is Yolwumne 91X-3 with measured depth of 14,300 feet. However, the well is surpassed only two years later by the relief well for the Bellevue blowout.
- 1998 - A blowout and oil well fire at the Bellevue #1 wildcat in the East Lost Hills subthrust fuels hopes for the first major Kern County discovery in over a decade. However, subsequent drilling proves to be a disappointment.

The Oil Industry of Medieval Persia (Azerbaijan and Baku)

When Marco Polo in 1264 visited the Persian city of Baku, on the shores of the Caspian Sea in modern Azerbaijan, he saw oil being collected from seeps. He wrote that "on the confines toward Geirgine there is a fountain from which oil springs in great abundance, inasmuch as a hundred shiploads might be taken from it at one time." In addition to oil seeps, Marco Polo also saw spectacular mud volcanos, sourced by natural gas seeping through ponds, and a flaming hillside, the "Eternal Fires of the Apsheron Peninsula", a spit of land that juts eastward from Azerbaijan into the Caspian Sea and where condensate and natural gas seeping through fractured shales has burned, and been worshipped, for centuries. The thumbnail on the right shows the Temple of the Fire Worshipers at Ateshkah where a gas seep has burned since ancient times.

Shallow pits were dug at the Baku seeps in ancient times to facilitate collecting oil, and hand-dug holes up to 35 metres (115 feet) deep were in use by 1594. These holes were essentially oil wells, which makes Baku the first true field. Apparently 116 of these wells in 1830 produced 3,840 metric tons (about 710 to 720 barrels) of oil. Later, Russian engineer F.N. Semyenov used a cable tool in 1844 to drill an oil well near the Bibi-Eibat (Bibi-Heybat) embayment on the Apsheron Peninsula, ten years before Colonel Drake's famous well in Pennsylvania.

Also, offshore drilling started up at Baku at Bibi-Eibat near the end of the 19th century, about the same time that the "first" offshore oil well was drilled in 1896 at Summerland field on the California Coast. The thumbnail on the left shows workers digging an oil well by hand at Bibi-Eibat. Baku was known for spectacular gushers, and spectacular well fires as well - an example of which is shown in the still on the right from a short film by the Lumière Brothers in 1896. The first of the big spouters blew out in 1873 on the Balakhani Plateau, which was the high ground of the Apsheron Peninisula, beneath which lies a giant anticline that is responsible for the prolific oil production. The Balakhani field during the 1870s was the largest oil field in the world. Another giant anticline at the Bibi-Eibat embayment, on the south side of the peninsula, extends from the land out to beneath the waters of the bay.

A string of huge onshore producers, beginning with the Tagiev spouter in the mid-1880s, elevated Bibi-Eibat to the largest field. Russian engineers, realising that production extended offshore, began filling and draining the bay in 1909 to allow continued drilling. Over

300 hectares (741 acres) had been reclaimed by 1927, a project said to be second in magnitude only to construction of the Panama Canal.

The Early Oil Industry of Poland and Romania

The Carpathian Mountains in Poland abound in oil seeps, and Carpathian oil, hand dipped from pits dug in front of the seeps, was burned in street lamps, as early as the 1500s, to provide light in the Polish town of Krosno. Unfortunately, the seep oil was a dark, viscous liquid that stuck to everything. It also burned with a foul smell and gave off more smoke and soot than other lamp oils, most of which were rendered from animal fat.

Ignacy Lukasiewicz, a Polish druggist in the modern Ukranian town of Lvov, saw the potential of using seep oil in lamps as a cheap alternative to expensive whale oil. To make a clean-burning fuel, he began experimenting with distillation techniques, perfected earlier by Dr. Abraham Gesner in Canada, to produce clear kerosene from smelly seep oil. His experiments gained notoriety, and the European oil industry was born on a dark night on July 31, 1853 when Lukasiewicz was called to a local hospital to provide light from one of his lamps for an emergency surgery. Impressed with his invention, the hospital ordered several lamps and 500 kg of kerosene. Lukasiewicz enlisted the aid of a business partner and travelled to the Vienna, capitol city of the Austro-Hungarian Empire, to register his distillation process with the government on December 31, 1853. To provide oil for his kerosene business, Lukasiewicz initially collected a thick, sticky crude from shallow, hand-dug wells in the Gorlice region, an area in the Carpathians about 50 miles west of the Polish town of Bóbrka. The following year, he teamed up with Titus Trzecieski and Mikolaj Klobassa to establish an "oil mine" in Bóbrka which pumped crude oil from hand-drilled, 30- to 50-metre deep wells. Later, wells as deep as 150 metres were drilled that produced a lighter, better-quality crude from which to distill kerosene. Other entrepreneurs dug their own wells, and a thriving Polish oil industry developed, which was followed in 1857 by the drilling of wells at Bend, northeast of Bucharest, on the Romanian side of the Carpathians. Two years later, Colonel Edwin Drake, who perhaps had knowledge of the Polish developments, drilled his famous well in Pennsylvania, an event wrongly labelled by many in the industry as the drilling of the "first oil well".

Many of these early wells were labouriously dug by hand. Others were drilled with spring poles, in which a springy wooden pole was stuck in the ground at an angle and a heavy metal drill bit attached by a cable to the head of the pole. Operators would bounce up and down on stirrups attached to the pole, causing the bit to literally chop a hole into the hard ground. The hole was cleaned by lowering into the hole a specially designed bucket, called a bailer, which was similarly bounced up and down until it filled dirt and cuttings to be hauled to the surface.

Steam engines were employed to mechanically drill wells in the Pennsylvania oil fields during the U.S. Civil War, and Thomas Bard imported a steam-powered drilling rig and crew from Pennsylvania to successfully drill a mediocre oil well in California in 1865. Steam was first used in Poland two years later in 1867 to drill a well at Kleczany, 60 kilometres west of the Bóbrka field. Steam-powered drilling made its debut at Bóbrka a few years later, sometime between 1870 and 1872, and enabled operators to drill much deeper than they had been able to previously. Within a few years virtually all oil wells, in both the United States and Europe, were being drilled mechanically.

The Early Oil Industry of Pennsylvania

Oil Creek in western Pennsylvania abounds in oil seeps that ooze thick black crude into the stream. These seeps were well known to the Seneca Indians, one of the Iroquois Nation tribes, who used the oil as a salve, mosquito repellent, purge and tonic. Many settlers also believed that these oils were medicinal, and "hawkers" sold bottles of it, as early as 1792, as a cure-all called "Seneca Oil". The nearby Allegheny and Kiskiminetas river valleys had oil also, but beneath the ground, where as early as 1815 it was contaminating several of the brine wells that supplied a booming salt industry in the Pittsburgh area.

In the early 1850s, a Pittsburgh druggist named Samuel Kier began selling bottled oil from his father's brine wells as "Pennsylvania Rock Oil", but met with little success. One day, Colonel A. C. Ferris, a whale oil dealer, processed a small amount of Kier's "tonic" to make a lighter oil that burned well in a lamp. When Kier heard about this, he began using a one-barrel whiskey still of his own to convert his rock oil into lamp oil. After Kier upgraded his still to five-barrel capacity, Pittsburgh forced him to move his operation to a suburb out of fear of an explosion.

When George Bissell, a New York lawyer, learned of Kier's operation, he hired Benjamin Silliman Jr of Yale University, probably around 1854, to see if Seneca Oil would yield lamp oil. Silliman successfully distilled the oil into several fractions, including an illuminating oil already known as kerosene. Armed with Silliman's results, Bissell received financial backing to form the "Pennsylvania Rock Oil Company", which later became the "Seneca Oil Company".

An unemployed railroad conductor and express agent named Edwin Drake, who by chance was staying at the same hotel in New Haven,

Connecticut as Bissel and his partners, was hired in 1857 to visit Titusville, a town on Oil Creek. Drake's only qualification for this assignment was a free railroad pass remaining from his previous job. Although Drake had never been in the military, when he returned to Titusville the following year to commence operations as a Seneca Oil Company agent, his employers passed him off as a colonel to give their venture an air of respectability.

Historically, oil was collected at Oil Creek by damming the creek near a seep, then skimming oil off the top of the resulting pond. Drake tried this at a seep once used by a sawmill to produce oil for lubricating the mill machinery, but even with improvements and opening up other seeps in the area, he only increased production from three or four gallons to a still non-economic six to ten gallons a day. Next workers tried digging a shaft to mine the oil, but groundwater flooded in too quickly for the workers to continue. Finally, Drake decided to drill a well and locate the source of the seep oil, using the same steam-powered equipment used to drill brine wells.

He hired a blacksmith named "Billy" Smith, who had drilled brine wells for Kier and others in the Pittsburgh area. Smith, with his son Samuel, began drilling in the summer of 1859. Although progress was slow, usually three feet a day in shale bedrock, they reached a depth of 69½ feet by August 27, just as Drake was reaching the last of his funds. When Billy and Samuel pulled their drilling tools from the well the next morning, they noticed oil rising in the hole. After installing a hand-operated lever pump borrowed from a local kitchen, the first days production was about twenty-five barrels. Production soon dropped off to a steady ten barrels or so a day, and the well is said to have continued at that rate for a year or more.

Although Drake's well was no gusher, it was the beginning of an idea. Titusville transformed almost overnight from a quiet farm town to an oil boom town of muddy roads, hastily constructed wooden derricks, and noisy steam engines. The Pennsylvania oil boom was on.

Spindletop, Texas

The modern oil industry was born on a hill in southeastern Texas. This hill was formed by a giant underground dome of salt as it moved slowly towards the surface. As it crept, it pushed the earth that was in its path higher and higher. This dome was known by several names, but the one that stuck was "Spindletop". Through the later half of the 19th century, Pennsylvania had been the most oil-productive state in the country. All that changed on January 10th, 1901.

Native American Indians in the area had been aware of oil seeps for centuries, and used this tar they found at the surface to treat a variety of ailments. Some would even drink the stuff in hopes that it could cure digestive problems. In 1543, Spanish explorers discovered that black, sticky tar found washed up on the beaches along the Texas coast could be used to waterproof their boots.

In the late 1800's, Texas had produced minor amounts of oil, starting with a well in 1866 drilled by Lyne T. Barret near the east Texas town of Nacogdoches. This field, known as "Oil Springs", was finally exploited again in 1888, when a crew of drillers from Pennsylvania had a well come in at 250 - 300 barels per day. This find attracted other oil companies, and it would only be a matter of time before the huge, untapped potential of the underground reservoirs was discovered. Corsicana was really the first big producing field in eastern Texas. This field, like so many others in the early days of oil, was discovered when local businessmen drilled deep wells looking for water, not oil. Amazingly, the first drillers in this area perceived the oil zones they reached as annoying nuisances, and often drilled past these zones to get to the water they were seeking. H.G.Damon and Ralph Beaton had a bit more foresight, and formed the Corsicana Oil Development Company. They brought in famed Pennsylvania oilman John Galey, and the team drilled marginally successful oil wells in Corsicana in 1896. These wells, however, all flowed at 25 barrels per day or less, and Galey and his partner, James Guffey, sold their interest in the company and headed back east, convinced that there was little future in Texas oil. Locals, however, proved them wrong, and by the end of the year 1900, more than 2 million barrels of oil had been produced in the Corsicana field alone. This was not large by Pennsylvania standards, but it certainly pointed towards the possibilities of something bigger.

Patillo Higgins, a one-armed mechanic and self-taught geologist, was one of the few at the time who believed that, in the future, modern industry would switch from coal to oil. But where to get all that oil? He believed it lay beneath his feet at Spindletop. He had a feeling that drilling a well on top of this salt dome (and others like it) would produce oil, and lots of it. In an attempt to turn his dream into a reality, Higgins organised the Gladys City Oil, Gas, and Manufacturing Company in 1892. Years of frustration followed, with most members of the petroleum and geologic communities proclaiming Higgins's ideas to be silly nonsense. Nearing the end of his rope, Higgins ran an advertisement in a local newspaper, and one man, Captain Anthony F. Lucas, replied.

Captain Lucas, unlike his predecessor in Pennsylvania, "Colonel" Edwin Drake, was a real Captain, having served in the Austrian Navy.

Lucas had training as an engineer and experience as a salt miner in Louisiana. But his first wells drilled for Higgins were failures, and the money ran out. The Texas press, as well as the local geologists, had been very skeptical of Higgins for years, and no one in the area believed that a salt dome structure could produce oil. So, Lucas turned to Guffey and Galey, who had left the area 3 years earlier, unconvinced of the potential of Texas oil. Something made them change their minds, and in 1900, John Galey returned to Beaumont, Texas to survey the area. He picked the spot, and the drilling began on October 27, 1900.

Drilling was difficult at first. Lucas and his men ran into the same problems that other drillers had faced along the Texas coastal plain. There is little in the way of rock at the surface in that part of the world. Instead, oil wildcatters had to drill through several hundred feet of sand. This made the hole prone to cave in on them. To help solve this problem, one of Lucas's drillers, Curt Hamill, came up with a solution that was revolutionary at the time. Instead of pumping water down the hole to flush out the cuttings produced by the action of the drill, he used mud. This proved to help not only in retrieving the cuttings, but just as importantly, it was found that the mud stuck to the sides of the hole and kept it from caving in. It was found there were even more benefits, and mud has been used in almost every drillhole around the world ever since.

Exhausted after 2 months of strenuous drilling, the crew shut down for a week over the Christmas holiday, 1900, having reached a depth of 880 feet. When they arrived back on site on New Year's Day, 1901, they were energized, and within a week drilled down to a depth of 1,020 feet. After pulling the drill out to change some equipment, they started to lower it back in. The day was January 10, 1901. After lowering it back into the open hole to a depth of about 700 feet, a full 17 hours since the last "drilling" had actually occurred, mud started bubbling back up the hole. Seconds later, the drill pipe shot out of the ground with great force, and then...nothing happened.

After a short time, the frustrated and confused drillers set about to clean up the mess and see if anything could be salvaged. All of a sudden, a noise like a cannon shot came from the hole, and mud came shooting out of the ground like a rocket. Within a few seconds, natural gas, then oil followed. The oil "gusher" - greenish-black in colour, doubled the size of the drilling derrick, rising to a height of more than 150 feet (about 50 metres). This was more oil than had ever been seen anywhere in the entire world. Captain Lucas had been hopeful that this well might produce 5 barrels per day. In fact, this well, "Lucas 1", flowed at an initial rate of nearly 100,000 barrels per day, more than all of the other producing wells in the United States COMBINED!

There was much money to be made at Spindletop, but there was even more money to lose. It is estimated that $50 million dollars was made from Spindletop, compared with investments equalling $80 million. As had happened in other booms, there were many frauds and cheats, and Spindletop was renamed "Swindletop" by some of the less fortunate and more cynical locals. The population of Beaumont swelled from 10,000 to 50,000 in a matter of months. Before the end of the year, Lucas's well was joined by over 200 other wells, all fighting for space on top of Spindletop. These wells were owned by more than 100 different oil companies. The gusher at Spindletop was responsible for creating several companies that were to become giants in the oil industry, including Gulf Oil, Amoco and Humble Oil Company, which would later become an arm of Exxon.

Now, people realised the true potential of oil - the vast resources that were, in fact, there, and the vast potential that remained. Before Spindletop, oil was used many for lamps and lubrication - after Spindletop, petroleum would be used as a major fuel for such new inventions as the airplane and automobile. Ships and trains that had previously run on the power of coal, now began to switch to oil, becoming convinced that there would be no shortage of the fuel anytime soon. As an example, the Santa Fe Railroad went from only one oil-driven locomotive in 1901 to two hundred and twenty-seven in 1905.

It is difficult to overstate the importance of Spindletop. From this point on, nothing in the oil industry would ever be the same again, for this "gusher" ushered in the modern age of petroleum.

The Story of Oil in California

Many people may be surprised to learn that one of Southern California's chief exports over the last 100 years, besides motion pictures, has been oil. Like oil reservoirs in Texas, Louisiana, and Pennsylvania, a hint of what lay beneath the surface could be seen in the many above ground oil seeps. These seeps had been known by Native Americans for thousands of years. In 1543, Spanish explorer Juan Cabrillo noticed the native people using the naturally occurring tar, or "pitch", to waterproof their canoes. These seeps were also responsible for the tar pits of La Brea (Spanish for "pitch"), which had, over many thousands of years, trapped unsuspecting animals and their predators looking for an easy meal. In 1865, only 6 years after "Colonel" Edwin Drake's monumental discovery in Pennsylvania, California's first productive well was drilled by the Union Matolle Company in California's Central Valley. This area, east of San Francisco, became the scene of much of the drilling activity through the rest of the 1800's.

While none of these wells were considered major strikes, they did provide enough oil for the nearby market of San Francisco, by far the largest population centre in California in the late 1800's.

It Came From Over There

But the largest fields lay undiscovered, near the sleepy seaside village of Los Angeles. The first well to strike oil in Southern California was drilled in 1892 by Edward L. Doheny, an unsuccessful gold and silver prospector, and Charles A. Canfield, his old mining partner. According to legend, Edward L. Doheny was in the downtown area of Los Angeles when he saw a cart whose wheels were coated in tar. When he asked the man where the substance had come from, he pointed to the northeast. Doheny and Canfield examined the area and soon discovered the Los Angeles Field after drilling to a depth of 140 metres (460 feet) at the corner of Colton Street and Glendale Boulevard, near present day Dodger Stadium. It was drilled using the unlikeliest of instruments: a sharpened end of a eucalyptus tree. Within 2 years of the find, 80 wells were producing oil in the area bounded by Figueoa, First, Union and Temple Streets. By 1897, the number of wells increased to 500.

Rise to Fame

Doheny would eventualy become a millionaire, and gain enough renown to challenge for the Democratic nomination for Vice-President of the United States in 1920. And although he was cleared of any wrong-doing, he would later become a central figure in the Teapot Dome Scandal of the 1920's which brought disgrace to the presidency of Warren G.

The Oil Queen

A local music teacher, Emma Summers, was one of the most successful investors in the first years of the initial boom, and by 1900, Summers controlled half the production in the original Los Angeles Field. For obvious reasons, Summers became known as "California's Petroleum Queen."

The oil boom in the early days attracted some interesting characters, including prostitutes, gamblers and con-men. The population of the city of Los Angeles doubled between 1890 and 1900, then tripled again between 1900 and 1910. Later, wells in the 1930's and 40's were soundproofed with vinyl-coated glass cloth with one-inch sheet fibreglass filling to decrease the noise, as the drilling activity began to conflict with the exploding Los Angeles population. Camouflage was also used, a technique that was eventually moved to offshore fields as well. In 1900, the state of California produced 4 million barrels. In 1910, this had jumped to 77 million barrels. In spite of this increased production, many of the fields were beginning to see slowdowns in their production rates in the late 1910's, and California's wondered if their oil boom was reaching an end. But before that would happen, 3 major fields were discovered in rapid succession - Huntington Beach (1920), Santa Fe Springs (1921), and the biggest of them all, the Signal Hill, or Long Beach, Field in 1921.

Signal Hill

Signal Hill rises up 110 metres (365 feet) behind Long Beach, 32 km (20 miles) south of Los Angeles. Its name is derived from a local Native American practice of signalling to each other from the imposing hill. Because of its size, signals could be sent by way of smoke or fire either to other hills in the area, or to boats out at sea. Oil men first started exploring the area in 1916 after the successes of other ventures in southern California. In 1921, Dr. W. Van Holst Pellekaan, Chief geologist for Shell, tried to stop the drilling at Signal Hill, unconvinced of its potential. He was too late, however, and the drilling proceeded.

The Shell Game

Shell's reluctance to drill Signal Hill was understandable. The company had spent three million dollars at Ventura in the previous 5 years, and had no oil to show for it. And only 4 years before, Union Oil had drilled an unsuccessful well (also known as a "duster") on Signal Hill. But it was ultimately the tenacity of Frank Hayes and Alvin Theodore Schwennesen, geologists with Shell, that moved the project forward.

Work began on the Alamitos # 1 well on March 23rd. By May 2, the hole reached 843 metres (2,765 feet) and gave a showing of oil. Soon thereafter, 21 metres (70 feet) of standing oil was found in the bottom of the hole. But still, no oil flowed, and the crew, lead by driller O.P. "Happy" Yowells, began to wonder what exactly was happening. Then on June 23rd at 9:30 PM, the Alamitos #1 erupted with so great a gas pressure that oil gushed 35 metres (114 feet) into the air. Unfortunately, the bottom of the hole soon caved in. Much cleaning of the hole was required, and on June 25, 1921, the well was producing more than 1,000 barrels of oil per day. The well would eventually produce 700,000 barrels of oil.

The rush is on....

The discovery created a stampede. While the well was being drilled, the area was in the process of being subdivided into residential lots. Many of the lots, though already sold to prospective homeowners, were not yet built upon, and potential homeowners quickly changed their minds and entered the business of looking for oil, hoping to get rich quick.

The parcels of land were so small and the forest of tall wooden derricks so thick that the legs of many of them actually intertwined. Oil promoters were selling shares of wells that had not yet been drilled. Signal Hill was to prove so prolific that, almost unbelievably, many of those buyers actually made money on their investments. The next-of-kin of persons buried in the Sunnyside Cemetary on Willow Street would eventually receive royalty checks for oil drawn out from beneath family grave plots.

By April 1922, only 10 months after completion of the discovery well, Signal Hill was covered with 108 wells, producing 14,000 barrels daily. By the fall of 1923, 259,000 barrels of crude was being produced every day from nearly 300 wells.

Signal Hill was the biggest field the already productive Southern California region had ever seen. In 1923, Signal Hill produced 244,000 barrels, alongside Huntington Beach (discovered in 1920) at 113,000 and Santa Fe (1921) at 32,000. This made California the nation's number-one producing state, and in 1923, California was the source of one-quarter of the world's entire output of oil! Even so, fears of shortage were still very much in the air.

"The supply of crude petroleum in this country is being rapidly depleted", the Federal Trade Commission warned in 1923. But in that same year, American crude oil production exceeded domestic demand for the first time in a decade.

Remote Location Turns into Tanker Technology

Because of California's remote location relative to the industrial centres of the east, California oil companies were at the forefront of tanker technologies. As a consequence, much of the state's market was overseas. In 1894, the Pacific Coast Oil Company and the Union Oil Company partnered to build the first true oil tanker on the Pacific Ocean. Named the George Loomis, its maiden voyage departed Ventura, California in January, 1896, and a new era was born. The development of California oil also presented challenges to the geologist that had been seen in no other oil field. As a result, the complexities of the geology of Southern California lead to a significantly increased knowledge of petroleum geology and exploration.

By the end of 1938, the Long Beach Field had produced 614.5 million barrels of crude, 750 million barrels by 1950, and over 900 million barrels by 1980. This made Signal Hill one of the most productive fields per acre the world has ever known.

Oil Well

An oil well is a general term for any boring through the earth's surface that is designed to find and acquire petroleum oil hydrocarbons. Usually some natural gas is produced along with the oil. A well that is designed to produce mainly or only gas may be termed a gas well.

History

The earliest known oil wells were drilled in China in 347 CE. They had depths of up to about 800 feet (240 m) and were drilled using bits attached to bamboo poles. The oil was burned to evaporate brine and produce salt. By the 10th century, extensive bamboo pipelines connected oil wells with salt springs. The ancient records of China and Japan are said to contain many allusions to the use of natural gas for lighting and heating. Petroleum was known as *burning water* in Japan in the 7th century. The Middle East's petroleum industry was established by the 8th century, when the streets of the newly constructed Baghdad were paved with tar, derived from petroleum that became accessible from natural fields in the region. Petroleum was distilled by the Persian alchemist Muhammad ibn Zakarîya Râzi (Rhazes) in the 9th century, producing chemicals such as kerosene in the alembic (*al-ambiq*), and which was mainly used for kerosene lamps. Arab and Persian chemists also distilled crude oil in order to produce flammable products for military purposes.

Some sources claim that from the 9th century, oil fields were exploited in the area around modern Baku, Azerbaijan, to produce naphtha for the petroleum industry. These fields were described by Marco Polo in the 13th century, who described the output of those oil wells as hundreds of shiploads. When Marco Polo in 1264 visited the Azerbaijani city of Baku, on the shores of the Caspian Sea, he saw oil being collected from seeps. He wrote that "on the confines toward Geirgine there is a fountain from which oil springs in great abundance, in as much as a hundred shiploads might be taken from it at one time."

Shallow pits were dug at the Baku seeps in ancient times to facilitate collecting oil, and hand-dug holes up to 35 metres (115 ft) deep were in use by 1594. These holes were essentially oil wells. Apparently 116 of these wells in 1830 produced 3,840 metric tons (about 28,000 barrels) of oil. Also, offshore drilling started up at Baku (then Russian Empire) at Bibi-Eibat field in 1846. In the New World, the first commercial oil well entered operation in Oil Springs, Ontario in 1858, while the first offshore oil well was drilled in 1896 at the Summerland Oil Field on the California Coast.

The earliest oil wells in modern times were drilled percussively, by hammering a cable tool into the earth. Soon after, cable tools were replaced with rotary drilling, which could drill boreholes to much greater depths and in less time. The record-depth Kola Borehole used non-rotary mud motor drilling to achieve a depth of over 12,000 metres (39,000 ft). Until the 1970s, most oil wells were vertical, although lithological and mechanical imperfections cause most wells to deviate at least slightly from true vertical. However, modern directional drilling technologies allow for strongly deviated wells which can, given sufficient depth and with the proper tools, actually become horizontal. This is of great value as the reservoir rocks which contain hydrocarbons are usually horizontal, or sub-horizontal; a horizontal wellbore placed in a production zone has more surface area in the production zone than a vertical well, resulting in a higher production rate.

The use of deviated and horizontal drilling has also made it possible to reach reservoirs several kilometres or miles away from the drilling location (extended reach drilling), allowing for the production of hydrocarbons located below locations that are either difficult to place a drilling rig on, environmentally sensitive, or populated.

Life of a Well

The creation and life of a well can be divided up into five segments:

- Planning
- Drilling
- Completion
- Production
- Abandonment

Drilling

The well is created by drilling a hole 5 to 50 inches (127.0 mm to 914.4 mm) in diametre into the earth with a drilling rig that rotates a drill string with a bit attached. After the hole is drilled, sections of steel pipe (casing), slightly smaller in diametre than the borehole, are placed in the hole. Cement may be placed between the outside of the casing and the borehole.

The casing provides structural integrity to the newly drilled wellbore, in addition to isolating potentially dangerous high pressure zones from each other and from the surface.

With these zones safely isolated and the formation protected by the casing, the well can be drilled deeper (into potentially more-unstable and violent formations) with a smaller bit, and also cased with a smaller size casing. Modern wells often have two to five sets of subsequently smaller hole sizes drilled inside one another, each cemented with casing.

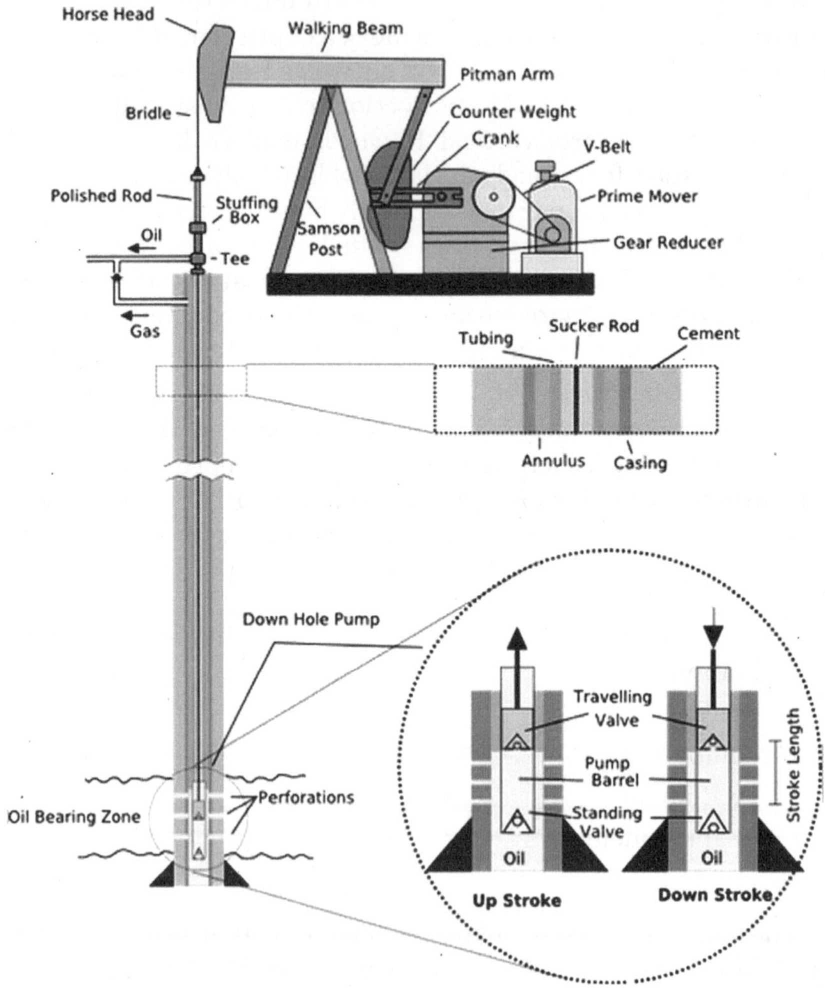

***Figure:** A schematic of a typical oil well being produced by a pumpjack, which is used to produce the remaining recoverable oil after natural pressure is no longer sufficient to raise oil to the surface.*

To drill the well:

- The drill bit, aided by the weight of thick walled pipes called "drill collars" above it, cuts into the rock. There are different types of drill bit; some cause the rock to disintegrate by compressive failure, while others shear slices off the rock as the bit turns.

- Drilling fluid, a.k.a. "mud", is pumped down the inside of the drill pipe and exits at the drill bit. Drilling mud is a complex mixture of fluids, solids and chemicals that must be carefully tailored to provide the correct physical and chemical characteristics required to safely drill the well. Particular functions of the drilling mud include cooling the bit, lifting rock cuttings to the surface, preventing destabilisation of the rock in the wellbore walls and overcoming the pressure of fluids inside the rock so that these fluids do not enter the wellbore.
- The generated rock "cuttings" are swept up by the drilling fluid as it circulates back to surface outside the drill pipe. The fluid then goes through "shakers" which strain the cuttings from the good fluid which is returned to the pit. Watching for abnormalities in the returning cuttings and monitoring pit volume or rate of returning fluid are imperative to catch "kicks" early. A "kick" is when the formation pressure at the depth of the bit is more than the hydrostatic head of the mud above, which if not controlled temporarily by closing the blowout preventers and ultimately by increasing the density of the drilling fluid would allow formation fluids and mud to come up through the drill pipe uncontrollably.
- The pipe or drill string to which the bit is attached is gradually lengthened as the well gets deeper by screwing in additional 30-foot (9 m) sections or "joints" of pipe under the kelly or topdrive at the surface. This process is called making a connection. Usually, joints are combined into three joints equalling one stand. Some smaller rigs only use two joints and some rigs can handle stands of four joints.

This process is all facilitated by a drilling rig which contains all necessary equipment to circulate the drilling fluid, hoist and turn the pipe, control downhole, remove cuttings from the drilling fluid, and generate on-site power for these operations.

Completion (Oil and Gas Wells)

In petroleum production, completion is the process of making a well ready for production (or injection). This principally involves preparing the bottom of the hole to the required specifications, running in the production tubing and its associated down hole tools as well as perforating and stimulating as required. Sometimes, the process of running in and cementing the casing is also included.

Lower Completion

This refers to the portion of the well across the production or injection zone. The well designer has many tools and options available to design the lower completion according to the conditions of the reservoir. Typically, the lower completion is set across the productive zone using a liner hanger system, which anchors the lower completion to the production casing string.

Barefoot Completion

Selective isolation of oil, gas and barefoot section. Its an open hole, but with simple design and low cost. can be used in consolidated formations. It's the technique used by drilling companies in deep wells, Natural fractured reservoirs and Multilateral.

Open Hole Completion

This designation refers to a range of completions where no casing or liner is cemented in place across the production zone. In competent formations the zone might be left entirely bare, but usually some sort of sand-control and/or flow-control means are incorporated. Openhole completions have seen significant uptake in recent years, and there are many configurations, often developed to address specific reservoir challenges. There have been many recent developments that have boosted the success of openhole completions, and they also tend to be popular in horizontal wells, where cemented installations are more expensive and technically more difficult. The common options for openhole completions are;

1) pre-holed liner (also often called pre-drilled liner). The liner is prepared with multiple small drilled holes, then set across the production zone to provide wellbore stability and an intervention conduit. Pre-holed liner is often combined with openhole packers, such as swelling elastomers, mechanical packers or external casing packers, to provide zonal segregation and isolation. It is now quite common to see a combination of pre-holed liner, solid liner and swelling elastomer packers to provide an initial isolation of unwanted water or gas zones. Multiple sliding sleeves can also be used in conjunction with openhole packers to provide considerable flexibility in zonal flow control for the life of the wellbore.

This type of completion is also being adopted in some water injection wells, although these require a much greater performance envelope for openhole packers, due to the considerable pressure and temperature changes that occur in water injectors.

Openhole completions (in comparison with cemented pipe) require better understanding of formation damage, wellbore clean-up and fluid loss control. A key difference is that perforating penetrates through the first 6-18 inches (15-45 cm) of formation around the wellbore, whilst openhole completions require the reservoir fluids to flow through all of the filtrate-invaded zone around the wellbore and lift-off of the mud filter cake.

Many openhole completions will incorporate fluid loss valves at the top of the liner to provide well control whilst the upper completion is run.

There are an increasing number of ideas coming into the market place to extend the options for openhole completions; for example, electronics can be used to actuate a self-opening or self-closing liner valve. This might be used in an openhole completion to improve clean-up, by bringing the well onto production from the toe-end for 100 days, then self-opening the heel-end. Inflow control devices and intelligent completions are also installed as openhole completions.

Pre-holed liner may provide some basic control of solids production, where the wellbore is thought to fail in aggregated chunks of rubble, but it is not typically regarded as a sand control completion.

2) Slotted liner can be selected as an alternative to pre-holed liner, sometimes as a personal preference or from established practice on a field. It can also be selected to provide a low cost control of sand/solids production. The slotted liner is machined with multiple longitudinal slots, for example 2 mm x 50mm, spread across the length and circumference of each joint. Recent advances in laser cutting means that slotting can now be done much cheaper to much smaller slot widths and in some situation slotted liner is now used for the same functionality as sand control screens.

3) Openhole sand control is selected where the liner is required to mechanically hold-back the movement of formation sand. There are many variants of openhole sand control, the three popular choices being stand-alone screens, openhole gravel packs (also known as external gravel packs, where a sized sand 'gravel' is placed as an annulus around the sand control screen) and expandable screens. Screen designs are mainly wire-wrap or premium; wire-wrap screens use spiral-welded corrosion-resistant wire wrapped around a drilled basepipe to provide a consistent small helical gap (such as 0.012-inch (0.30 mm), termed 12 gauge). Premium screens use a woven metal cloth

wrapped around a basepipe. Expandable screens are run to depth before being mechanically swaged to a larger diametre.

Cased Hole Completion

This involves running casing or a liner down through the production zone, and cementing it in place. Connection between the well bore and the formation is made by perforating. Because perforation intervals can be precisely positioned, this type of completion affords good control of fluid flow, although it relies on the quality of the cement to prevent fluid flow behind the liner. As such it is the most common form of completion.

Completion Components

The upper completion refers to all components from the bottom of the production tubing upwards. Proper design of this "completion string" is essential to ensure the well can flow properly given the reservoir conditions and to permit any operations as are deemed necessary for enhancing production and safety.

Wellhead

A wellhead is a general term used to describe the component at the surface of an oil or gas well that provides the structural and pressure-containing interface for the drilling and production equipment.

The primary purpose of a wellhead is to provide the suspension point and pressure seals for the casing strings that run from the bottom of the hole sections to the surface pressure control equipment. While drilling the oil well, surface pressure control is provided by a blowout preventer (BOP). If the pressure is not contained during drilling operations by the column of drilling fluid, casings, wellhead, and BOP, a well blowout could occur. Once the well has been drilled, it is *completed* to provide an interface with the reservoir rock and a tubular conduit for the well fluids. The surface pressure control is provided by a christmas tree, which is installed on top of the wellhead, with isolation valves and choke equipment to control the flow of well fluids during production. Wellheads are typically welded onto the first string of casing, which has been cemented in place during drilling operations, to form an integral structure of the well. In exploration wells that are later abandoned, the wellhead may be recovered for refurbishment and re use. Offshore, where a wellhead is located on the production platform it is called a *surface wellhead*, and if located beneath the water then it is referred to as a subsea wellhead or mudline wellhead.

Functions: A wellhead serves numerous functions, some of which are:

1. Provide a means of casing suspension. (Casing is the permanently installed pipe used to line the well hole for pressure containment and collapse prevention during the drilling phase).
2. Provides a means of tubing suspension. (Tubing is removable pipe installed in the well through which well fluids pass).
3. Provides a means of pressure sealing and isolation between casing at surface when many casing strings are used.
4. Provides pressure monitoring and pumping access to annuli between the different casing/tubing strings.
5. Provides a means of attaching a blowout preventer during drilling.
6. Provides a means of attaching a christmas tree for production operations.
7. Provides a reliable means of well access.
8. Provides a means of attaching a well pump,

Components

The primary components of a wellhead system are:

- casing head
- casing spools
- casing hangers
- packoffs (isolation) seals
- bowl protectors / wear bushings
- test plugs
- mudline suspension systems
- tubing heads
- tubing hangers
- tubing head adapters

Design Specification

The oil industry specifications for wellhead systems (materials, dimensions, test procedures and pressure ratings etc.) are :

1. API 6A Specification for Wellhead and Christmas Tree Equipment
2. ISO 10423 Wellhead and Christmas Tree Equipment.

Christmas Tree (Oil Well)

This article needs additional citations for verification. Please help improve this article by adding citations to reliable sources. Unsourced material may be challenged and removed. *(March 2011)*

In petroleum and natural gas extraction, a Christmas tree, or "tree", or "vampire deterrent", (not "wellhead" as sometimes incorrectly referred to) is an assembly of valves, spools, and fittings used for an oil well, gas well, water injection well, water disposal well, gas injection well, condensate well and other types of wells. It was named for its crude resemblance to a decorated tree.

Overview

Note that a tree and wellhead are separate pieces of equipment not to be mistaken as the same piece. A wellhead must be present in order to utilise a Christmas tree and a wellhead is used *without* a Christmas tree during drilling operations, and also for riser tie-back situations which would then have a tree included at riser top. Producing surface wells that require pumps (pump jacks, nodding donkeys, and so on) frequently do not utilise any tree due to no pressure containment requirement.

Tree complexity has increased over the last few decades. They are frequently manufactured from blocks of steel containing multiple valves rather than made from multiple flanged valves. This is especially true in subsea applications where the resemblance to Christmas trees no longer exists given the frame and support systems into which the main valve block is integrated.

It is common to identify the type of tree as either "subsea tree" or "surface tree". Each of these classifications has a number of variations. Examples of subsea include conventional, dual bore, mono bore, TFL (through flow line), horizontal, mudline, mudline horizontal, side valve, and TBT (through bore tree) trees. The primary function of a tree is to control the flow into or out of the well, usually oil or gas. A tree often provides numerous additional functions including chemical injection points, well intervention means, pressure relief means (such as annulus vent), tree and well monitoring points (such as pressure, temperature, corrosion, erosion, sand detection, flow rate, flow composition, valve and choke position feedback, connection points for devices such as down hole pressure and temperature transducer (DHPT). When the operator, well, and facilities are ready to produce and receive oil or gas, valves are opened and the release of the formation fluids is allowed to flow into and through a pipeline. The pipeline then leads to a processing facility, storage depot and/or other pipeline eventually leading to a refinery or distribution centre (for gas). Subsea wells and thus trees usually flow through flowlines to a fixed or floating production platform or to a storage vessel (known as a floating storage offloading vessel

(FSO), or floating processing unit (FPU), or floating production and offloading vessel or FPSO or other combination of structures).

A tree may also be used to control the injection of gas or water injection application on a producing or non-producing well in order to sustain economic "production" volumes of oil from other well(s) in the area (field). On producing wells, injection of chemicals or alcohols or oil distillates to prevent and or solve production problems (such as blockages) may be used. Functionality may be extended further by using the control system on a subsea tree to monitor, measure, and react to sensor outputs on the tree or even down the well bore.

The control system attached to the tree controls the downhole safety valve (scssv, dhsv, sssv) while the tree acts as an attachment and conduit means of the control system to the downhole safety valve. Christmas trees are used on both surface and subsea wells (current technical limits are up to around 3000 metres and working temperatures of -50°F to 350°F with a pressure of up to 15,000 psi). The deepest installed subsea tree is in the Gulf of Mexico at approximately 9,000 feet (2,700 m).

Valves

Subsea and surface trees have a large variety of valve configurations and combinations of manual and/or actuated (hydraulic or pneumatic) valves. Examples are identified in API Specifications 6A and 17D.

A basic surface tree consists of two or three manual valves (usually gate valves because of their strength). A typical sophisticated surface tree will have at least four or five valves, normally arranged in a crucifix type pattern (hence the endurance of the term "Christmas tree"). The two lower valves are called the master valves (upper and lower respectively) because they lie in the flow path, which well fluids must take to get to surface. The lower master valve will normally be manually operated, while the upper master valve is often hydraulically actuated, allowing it to be a means of well control while an actuated wing valve is normally the primary well remotely (from control room or control panel) controlled valve. Hydraulic tree wing valves are usually built to be fail safe closed, meaning they require active hydraulic pressure to stay open.

The right hand valve is often called the flow wing valve or the production wing valve, because it is in the flowpath the hydrocarbons take to production facilities (or the path water or gas will take from production to the well in the case of injection wells).

The left hand valve is often called the kill wing valve. It is primarily used for injection of fluids such as corrosion inhibitors or methanol to prevent hydrate formation. In the North Sea, it is called the non-active side arm (NASA). It is typically manually operated.

The valve at the top is called the swab valve and lies in the path used for well interventions like wireline and coiled tubing. For such operations, a lubricator is rigged up onto the top of the tree and the wire or coil is lowered through the lubricator, past the swab valve and into the well. This valve is typically manually operated.

Some trees have a second swab valve, the two arranged one on top of the other. The intention is to allow rigging down equipment from the top of the tree with the well flowing while still preserving the Two-barrier rule. With only a single swab valve, the upper master valve is usually closed to act as the second barrier, forcing the well to be shut in for a day during rig down operations. However, avoiding delaying production for a day is usually too small a gain to be worth the extra expense of a having a Christmas tree with a second swab valve.

Subsea trees are available in either vertical or horizontal configurations with further speciality available such as dual bore, monobore, concentric, drill-through, mudline, guidlineless or guideline. Subsea trees may range in size and weight from a few tons to approximately 70 tons for high pressure, deepwater (>3000 feet) guidelineless applications. Subsea trees contain many additional valves and accessories compared to Surface trees.

Typically a subsea tree would have a choke (permits control of flow), a floline connection interface (hub, flange or other connection), subsea control interface (direct hydraulic, electro hydraulic, or electric) and sensors for gathering data such as pressure, temperature, sand flow, erosion, multiPhase flow, single phase flow such as water or gas.

Tubing Hanger

A tubing hanger is a component used in the completion of oil and gas production wells. It is set in the tree or the wellhead and suspends the production tubing and/or casing. Sometimes it provides porting to allow the communication of hydraulic, electric and other downhole functions, as well as chemical injection. It also serves to seal-in the annulus and production areas.

Production Tubing

Production tubing is a tube used in a wellbore through which production fluids are produced (travel).

Production tubing is run into the drilled well after the casing is run and cemented in place. Production tubing protects wellbore casing from wear, tear, corrosion, and deposition of by-products, such as sand / silt, paraffins, and asphaltenes. Along with other components that constitute the production string, it provides a continuous bore from the production zone to the wellhead through which oil and gas can be produced. It is usually between five and ten centimetres in diametre and is held inside the casing through the use of expandable packing devices. Purpose and design of production tubing is to enable quick, efficient, and safe installation, removal and re-installation. If there is more than one zone of production in the well, up to four lines of production tubing can be run.

Downhole Safety Valve

A downhole safety valve refers to a component on an oil and gas well, which acts as a failsafe to prevent the uncontrolled release of reservoir fluids in the event of a worst case scenario surface disaster. It is almost always installed as a vital component on the completion.

Operation

These valves are commonly uni-directional flapper valves which open downwards, such that the flow of wellbore fluids try to push it shut, while pressure from surface pushes it open. This means that when closed, it will isolate the reservoir fluids from surface.

Most downhole safety valves are controlled hydraulically from surface, meaning they are opened using a hydraulic connection linked directly to a well control panel. When hydraulic pressure is applied down a control line, the hydraulic pressure forces a sleeve within the valve to slide downwards. This movement compresses a large spring and pushes the flapper downwards to open the valve. When hydraulic pressure is removed, the spring pushes the sleeve back up and causes the flapper to shut. In this way, it is failsafe and will isolate the wellbore in the event of a loss of the wellhead. The full designation for a typical valve is 'tubing retrievable, surface controlled, subsurface safety valve', abbreviated to TR-SCSSV.

Positioning

The location of the downhole safety valve within the completion is a precisely determined parameter intended to optimise safety. There are arguments against it either being too high or too low in the well and so the final depth is a compromise of all factors. MMS regulations state that the valve must be placed no less than 100' below the mudline.

Reasons to Keep it High

The further down the well the DHSV is located, the greater the potential inventory of hydrocarbons above it when closed. This means that in the event of loss of containment at surface, there is more fluid to be spilled causing environmental damage, or in the worst case, more fuel for a fire. Therefore, placing the valve higher limits this hazard.

Another reason relates to the hydraulic control line. Hydraulic pressure is required to keep the valve open as part of the failsafe design. However, if the valve is too far down the well, then the portion of the control line below the surface is larger. If it is too large, then the weight of the hydraulic fluid alone may apply sufficient pressure to keep the valve open, even with the loss of surface pressurisation.

Reasons to Keep it Low

As part of the role of the DHSV to isolate the surface from wellbore fluids, it is necessary for the valve to be positioned away from the well where it could potentially come to harm. This implies that it must be placed subsurface in all circumstances, i.e. in offshore wells, not above the seabed. There is also the risk of cratering in the event of a catastrophic loss of the topside facility. The valve is specifically placed below the maximum depth where cratering is expected to be a risk.

If there is a risk of methane hydrate (clathrate) plugs forming as the pressure changes through the valve due to Joule-Thomson cooling, then this is a reason to keep it low, where the rock is warmer than 30 C.

Deploying and Retrieving

Most downhole safety valves installed as part of the completion design are classed as "tubing retrievable". This means that they are installed as a component of the completion string and run in during completion. Retrieving the valve, should it malfunction, requires a workover. The full name for this most common type of downhole safety valve is a Tubing Retrievable Surface Controlled Sub-Surface Valve, shortened in completion diagrams to TRSCSSV.

If a tubing retrievable valve fails, rather than go to the expense of a workover, a "wireline retrievable" valve may be used instead. This type of valve can fit inside the production tubing and is deployed on wireline after the old valve has been straddled open.

Legal Requirement

The importance of DHSVs is undisputed. Graphic images of oil wells in Kuwait on fire after the First Gulf War after their wellheads were removed, demonstrate the perils of not using the components (at the time, they were deemed unnecessary because they were onshore wells). It is, however, not a direct legal requirement in many places. In the United Kingdom, no law mandates the use of DHSVs. However, the 1974 Health & Safety at Work Act requires that measures are taken to ensure that the uncontrolled release of wellbore fluids is prevented even in the worst case. The brilliance of the act is that it does not issue prescriptive guideline for how to achieve the goal of health and safety, but merely sets out the requirement that the goal be achieved. It is up to the oil companies to decide how to achieve it and DHSVs are an important component of that decision. As such, although not a legal requirement, it is company policy for many operators in the UKCS.

Issues

While the DHSV isolates the production tubing, a loss of integrity could allow wellbore fluid to bypass the valve and escape to surface through the annulus. For wells using gas lift, it is often a requirement to install a safety valve in the 'A' annulus of the well to ensure that the surface is protected from a loss of annulus containment. However, these valves are not as common and they are not necessarily installed at the same position in the well, meaning it is possible that fluids could snake their way around the valves to surface.

Annular Safety Valve: On wells with gas lift capability, many operators consider it prudent to install a valve, which will isolate the 'A' annulus for the same reasons a DHSV may be needed to isolate the production tubing in order to prevent the inventory of natural gas downhole from becoming a hazard as it became on Piper Alpha.

Side Pocket Mandrel

This is a welded/machined product which contains a 'side-pocket' alongside the main tubular conduit. The side pocket, typically 1" or 1½" diametre is designed to contain gas lift valve, which allows hydrocarbon gas from the 'A' annulus to be injected into the flow stream.

Submersible Pump

A submersible pump (or electric submersible pump (ESP)) is a device which has a hermetically sealed motor close-coupled to the pump body. The whole assembly is submerged in the fluid to be pumped. The main advantage of this type of pump is that it prevents pump cavitation, a problem associated with a high elevation difference between pump and the fluid surface. Submersible pumps push fluid to the surface as opposed to jet pumps having to pull fluids. Submersibles are more efficient than jet pumps.

Working Principle

The submersible pumps used in ESP installations are multistage centrifugal pumps operating in a vertical position. Although their constructional and operational features underwent a continuous evolution over the years, their basic operational principle remained the same. Produced liquids, after being subjected to great centrifugal forces caused by the high rotational speed of the impeller, lose their kinetic energy in the diffuser where a conversion of kinetic to pressure energy takes place. This is the main operational mechanism of radial and mixed flow pumps.

The pump shaft is connected to the gas separator or the protector by a mechanical coupling at the bottom of the pump. Well fluids enter the pump through an intake screen and are lifted by the pump stages. Other parts include the radial bearings (bushings) distributed along the length of the shaft providing radial support to the pump shaft turning at high rotational speeds. An optional thrust bearing takes up part of the axial forces arising in the pump but most of those forces are absorbed by the protector's thrust bearing.

Applications

Submersible pumps are found in many applications. Single stage pumps are used for drainage, sewage pumping, general industrial pumping and slurry pumping. They are also popular with aquarium filters. Multiple stage submersible pumps are typically lowered down a borehole and used for water abstraction, water wells and in oil wells.

Special attention to the type of ESP is required when using certain types of liquids. ESP's commonly used on board naval vessels cannot be used to dewater contaminated flooded spaces. These use a 440 volt A/C motor that operates a small centrifugal pump. It can also be used out of the water, taking suction with a 2-1/2 inch non-collapsible hose. The pumped liquid is circulated around the motor for cooling purposes. There is a possibility that the gasoline will leak into the pump causing a fire or destroying the pump, so hot water and flammable liquids should be avoided.

ESP Usage in Oil Wells

Submersible pumps are used in oil production to provide a relatively efficient form of "artificial lift", able to operate across a broad range of flow rates and depths. By decreasing the pressure at the bottom of the well (by lowering bottomhole flowing pressure, or increasing drawdown), significantly more oil can be produced from the well when compared with natural production. The pumps are typically electrically powered and referred to as Electrical Submersible Pumps (ESP).

ESP systems consist of both surface components (housed in the production facility, for example an oil platform) and sub-surface components (found in the well hole). Surface components include the motor controller (often a variable speed controller), surface cables and transformers. Subsurface components typically include the pump, motor, seal and cables. A gas separator is sometimes installed.

The pump itself is a multi-stage unit with the number of stages being determined by the operating requirements. Each stage consists of a driven impeller and a diffuser which directs flow to the next stage of the pump. Pumps come in diametres from 90mm (3.5 inches) to 254mm (10 inches) and vary between 1 metre (3 ft) and 8.7 metres (29 ft) in length. The motor used to drive the pump is typically a three phase, squirrel cage induction motor, with a nameplate power rating in the range 7.5 kW to 560 kW (at 60 Hz).

New varieties of ESP can include a water/oil separator which permits the water to be reinjected into the reservoir without the need to lift it to the surface. There are at least 15 brands of oilfield esps used throughout the world. Until recently, ESPs had been highly costly to install due to the requirement of an electric cable downhole. This cable had to be wrapped around jointed tubing and connected at each joint. New coiled tubing umbilicals allow for both the piping and electric cable to deployed with a single conventional coiled tubing unit.

The ESP system consists of a number of components that turn a staged series of centrifugal pumps to increase the pressure of the well fluid and push it to the surface. The energy to turn the pump comes from a high-voltage (3 to 5 kV) alternating-current source to drive a special motor that can work at high temperatures of up to 300 °F (149 °C) and high pressures of up to 5,000 psi (34 MPa), from deep wells of up to 12,000 feet (3.7 km) deep with high energy requirements of up to about 1000 horsepower (750 kW). ESPs have dramatically lower efficiencies with significant fractions of gas, greater than about 10% volume at the pump intake. Given their high rotational speed of up to 4000 rpm (67 Hz) and tight clearances, they are not very tolerant of solids such as sand.

Landing Nipple

This is a receptacle to receive wireline tools. It is also a useful marker for depths in the well, which can be difficult to accurately determine. Normally it is set close to the end of the tubing string to be able to isolate the same from the reservoir conditions, at any time during the producing life of the well.

Sliding Sleeve

A sliding sleeve is a standard component for the completion of an oil or gas well. Their main uses are to shut off flow from one or more reservoir zones or to regulate pressure between zones.

There are two main categories of sliding sleeves: open/close and choking. Open/close sleeves are shifted between a full open position and a closed position. They are used to shut off flow from a zone for economic reasons or to shut off a zone that is depleting or producing too much water. In multi-zone wells, they are used to regulate which zones to produce from and which ones to shut off. Mechanically actuated sleeves are simple and inexpensive but require actuation by a "lock," which must be run in the well on wireline or coiled tubing. Hydraulically actuated sleeves are more complicated but can be actuated from a small pump at surface. King sleeves can be used to regulate the pressure between two or more zones. They are also used to regulate the flow of fluid into a well during proppant fracturing or hydraulic fracturing operations. Choking sleeves are all hydraulically actuated and have a much more complex design than open/close sleeves.

A "sliding sleeve is also:

1. Part of a pencil for heavy-handed users.
2. Sliding sleeve supports the lead for precise lines and retracts as the pencil is used.
3. The sleeve is designed to protect the lead and minimise lead breakage.
4. Ideal for heavy-handed writers needing to use fine lead

Production Packer

A production packer is a standard component of the completion hardware of oil or gas wells used to provide a seal between the outside of the production tubing and the inside of the casing, liner, or wellbore wall.

Based on its primary use, packers can be divided into two main categories: production packers and service packers. Production packers are those that remain in the well during well production. Service packers are used temporarily during well service activities such as cement squeezing, acidizing, fracturing and well testing.

It is usually run in close to the bottom end of the production tubing and set at a point above the top perforations or sand screens. In wells with multiple reservoir zones, packers are used to isolate the perforations for each zone. In these situations, a sliding sleeve would be used to select which zone to produce. Packers may also be used to protect the casing from pressure and produced fluids, isolate sections of corroded casing, casing leaks or squeezed perforations, and isolate or temporarily abandon producing zones. In waterflooding developments in which water is injected into the reservoir, packers are used in injection wells to isolate the zones into which the water must be injected.

There are occasions in which running a packer may not be desirable. For example, high volume wells that are produced both up the tubing and annulus will not include a packer. Rod pumped wells are not normally run with packers because the associated gas is produced up the annulus. In general, well completions may not incorporate a packer when the annular space is used as a production conduit.

A production packer is designed to grip and seal against the casing ID. Gripping is accomplished with metal wedges called "slips." These components have sharpened, carburized teeth that dig into the metal of the casing. Sealing is accomplished with large, cylindrical rubber elements. In situations where the sealed pressure is very high (above 5,000 psi), metal rings are used on either side of the elements to prevent the rubber from extruding.

A packer is run in the casing on production tubing or wireline. Once the desired depth is reached, the slips and element must be expanded out to contact the casing. Axial loads are applied to push the slips up a ramp and to compress the element, causing it to expand outward. The axial loads are applied either hydraulically, mechanically, or with a slow burning chemical charge.

Most packers are "permanent" and require milling in order to remove them from the casing. The main advantages of permanent packers are lower cost and greater sealing and gripping capabilities.

In situations where a packer must be easily removed from the well, such as secondary recoveries, re-completions, or to change out the production tubing, a retrievable packer must be used. To unset the tool, either a metal ring is sheared or a sleeve is shifted to disengage connecting components. Retrievable packers have a more complicated design and generally lower sealing and gripping capabilities, but after removal and subsequent servicing, they can be reused.

Downhole Gauges

This is an electronic or fibre optic sensor to provide continuous monitoring of downhole pressure and temperature. Gauges use a 1/4" control line clamped onto the outside of the tubing string to provide an electrical or fibre optic communication to surface.

Perforated Joint

This is a length of tubing with holes punched into it. If used, it will normally be positioned below the packer and will offer an alternative entry path for reservoir fluids into the tubing in case the shoe becomes blocked, for example, by a stuck perforation gun.

Formation Isolation Valve

This component, placed towards the foot of the completion string, is used to provide two way isolation from the formation for completion operations without the need for kill weight fluids. Their use is sporadic as they do not enjoy the best reputation for reliability when it comes to opening them at the end of the completion process.

Centraliser

In highly deviated wells, this components may be included towards the foot of the completion. It consists of a large collar, which keeps the completion string centralised within the hole.

Wireline Entry Guide

This component is often installed at the end of the tubing (the shoe). It is intended to make pulling out wireline tools easier by offering a guiding surface for the toolstring to re-enter the tubing without getting caught on the side of the shoe.

- Schlumberger WEG (pdf) Last accessed: 15-Jun-2011

Perforating and Stimulating

In cased hole completions (the majority of wells), once the completion string is in place, the final stage is to make a connection between the wellbore and the formation. This is done by running perforation guns to blast holes in the casing or liner to make a connection. Modern perforations are made using shaped explosive charges, similar to the armour-penetrating charge used on antitank rockets (bazookas).

Sometimes once the well is fully completed, further stimulation is necessary to achieve the planned productivity. There are a number of stimulation techniques.

Acidizing

This involves the injection of chemicals to eat away at any skin damage, "cleaning up" the formation, thereby improving the flow of reservoir fluids. Acid can also be used to clean the wellbore of some scales that form from mineral laden produced water.

Fracturing

This means creating and extending fractures from the perforation tunnels deeper into the formation, increasing the surface area for formation fluids to flow into the well, as well as extending past any possible damage near the wellbore. This may be done by injecting fluids at high pressure (hydraulic fracturing), injecting fluids laced with round granular material (proppant fracturing), or using explosives to generate a high pressure and high speed gas flow (TNT or PETN up to 1,900,000 psi (13,000,000 kPa)) and (propellant stimulation up to 4,000 psi (28,000 kPa)).

Acidising and Fracturing (Combined Method)

This involves use of explosives and injection of chemicals to increase acid-rock contact.

Nitrogen Circulation

Sometimes, productivity may be hampered due to the residue of completion fluids, heavy brines, in the wellbore. This is particularly a problem in gas wells. In these cases, coiled tubing may be used to pump nitrogen at high pressure into the bottom of the borehole to circulate out the brine.

Production: The production stage is the most important stage of a well's life, when the oil and gas are produced. By this time, the oil rigs and workover rigs used to drill and complete the well have moved off the wellbore, and the top is usually outfitted with a collection of valves called a Christmas tree or Production trees. These valves regulate pressures, control flows, and allow access to the wellbore in case further completion work is needed. From the outlet valve of the production tree, the flow can be connected to a distribution network of pipelines and tanks to supply the product to refineries, natural gas compressor stations, or oil export terminals.

As long as the pressure in the reservoir remains high enough, the production tree is all that is required to produce the well. If the pressure depletes and it is considered economically viable, an artificial lift method mentioned in the completions section can be employed.

Workovers are often necessary in older wells, which may need smaller diametre tubing, scale or paraffin removal, acid matrix jobs, or completing new zones of interest in a shallower reservoir. Such remedial work can be performed using workover rigs – also known as *pulling units* or *completion rigs* – to pull and replace tubing, or by the use of well intervention techniques utilising coiled tubing. Depending on the type of lift system and wellhead a rod rig or flushby can be used to change a pump without pulling the tubing.

Enhanced recovery methods such as water flooding, steam flooding, or CO_2 flooding may be used to increase reservoir pressure and provide a "sweep" effect to push hydrocarbons out of the reservoir. Such methods require the use of injection wells (often chosen from old production wells in a carefully determined pattern), and are used when facing problems with reservoir pressure depletion, high oil viscosity, or can even be employed early in a field's life. In certain cases – depending on the reservoir's geomechanics – reservoir engineers may determine that ultimate recoverable oil may be increased by applying a waterflooding strategy early in the field's development rather than later. Such enhanced recovery techniques are often called "tertiary recovery".

Abandonment

A well is said to reach an "economic limit" when its production rate does not cover the expenses, including taxes.

The economic limit for oil and gas wells can be expressed using these formula:

Oil fields:

$$EL_{oil} = \frac{WI \times LOE}{NRI[P_o + (P_g \times GOR)/1{,}000] \times (1 - T)}$$

Gas fields:

$$EL_{gas} = \frac{WI \times LOE}{NRI[(P_o \times Y) + P_g] \times (1 - T)}$$

Where:

EL_{oil} is an oil well's economic limit in oil barrels per month (bbls/month).

EL_{gas} is a gas well's economic limit in thousand standard cubic feet per month (MSCF/month).

P_o,P_g are the current prices of oil and gas in dollars per barrels and dollars per MSCF respectively.

LOE is the lease operating expenses in dollars per well per month.

WI working interest, as a fraction.

NRI net revenue interest, as a fraction.

GOR gas/oil ratio as bbls/MSCF.

Y condensate yield as barrel/million standard cubic feet.

T production and severance taxes, as a fraction.

When the economic limit is raised, the life of the well is shortened and proven oil reserves are lost. Conversely, when the economic limit is lowered, the life of the well is lengthened.

When the economic limit is reached, the well becomes a liability and is abandoned. In this process, tubing is removed from the well and sections of well bore are filled with cement to isolate the flow path between gas and water zones from each other, as well as the surface. Completely filling the well bore with cement is costly and unnecessary. The surface around the wellhead is then excavated, and the wellhead and casing are cut off, a cap is welded in place and then buried.

At the economic limit there often is still a significant amount of unrecoverable oil left in the reservoir. It might be tempting to defer physical abandonment for an extended period of time, hoping that the oil price will go up or that new supplemental recovery techniques will be perfected. However, lease provisions and governmental regulations usually require quick abandonment; liability and tax concerns also may favour abandonment.

In theory an abandoned well can be reentered and restored to production (or converted to injection service for supplemental recovery or for downhole hydrocarbons storage), but reentry often proves to be difficult mechanically and not cost effective.

Types of Wells

Fossil-fuel wells come in many varieties. By produced fluid, there can be wells that produce oil, wells that produce oil *and* natural gas, or wells that *only* produce natural gas. Natural gas is almost always a byproduct of producing oil, since the small, light gas carbon chains come out of solution as they undergo pressure reduction from the reservoir to the surface, similar to uncapping a bottle of soda pop where the carbon dioxide effervesces. Unwanted natural gas can be a disposal problem at the well site. If there is not a market for natural gas near the wellhead it is virtually valueless since it must be piped to the end user. Until recently, such unwanted gas was burned off at the wellsite, but due to environmental concerns this practice is becoming less common. Often, unwanted (or 'stranded' gas without a market) gas is pumped back into the reservoir with an 'injection' well for disposal or repressurizing the producing formation. Another solution is to export the natural gas as a liquid.

Gas-to-liquid, (GTL) is a developing technology that converts stranded natural gas into synthetic gasoline, diesel or jet fuel through the Fischer-Tropsch process developed in World War II Germany. Such fuels can be transported through conventional pipelines and tankers to users. Proponents claim GTL fuels burn cleaner than comparable petroleum fuels. Most major international oil companies are in advanced development stages of GTL production, e.g. the 140,000 bbl/d (22,000 m^3/d) Pearl GTL plant in Qatar, scheduled to come online in 2011. In locations such as the United States with a high natural gas demand, pipelines are constructed to take the gas from the wellsite to the end consumer.

Figure: *A natural gas well in the southeast Lost Hills Field, California, US.*

Another obvious way to classify oil wells is by land or offshore wells. There is very little difference in the well itself. An offshore well targets a reservoir that happens to be underneath an ocean. Due to logistics, drilling an offshore well is far more costly than an onshore well. By far the most common type is the onshore well. These wells dot the Southern and Central Great Plains, Southwestern United States, and are the most common wells in the Middle East.

Another way to classify oil wells is by their purpose in contributing to the development of a resource. They can be characterised as:

- *production wells* are drilled primarily for producing oil or gas, once the producing structure and characteristics are determined
- *appraisal wells* are used to assess characteristics (such as flow rate) of a proven hydrocarbon accumulation

- *exploration wells* are drilled purely for exploratory (information gathering) purposes in a new area
- *wildcat wells* are those drilled outside of and not in the vicinity of known oil or gas fields.

At a producing well site, active wells may be further categorised as:

- *oil producers* producing predominantly liquid hydrocarbons, but mostly with some associated gas.
- *gas producers* producing almost entirely gaseous hydrocarbons.
- *water injectors* injecting water into the formation to maintain reservoir pressure or simply to dispose of water produced with the hydrocarbons because even after treatment, it would be too oily and too saline to be considered clean for dumping overboard, let alone into a fresh water source, in the case of onshore wells. Frequently water injection has an element of reservoir management and produced water disposal.
- *aquifer producers* intentionally producing reservoir water for re-injection to manage pressure. This is in effect moving reservoir water from where it is not as useful to where it is more useful. These wells will generally only be used if produced water from the oil or gas producers is insufficient for reservoir management purposes. Using aquifer produced water rather than water from other sources is to preclude chemical incompatibility that might lead to reservoir-plugging precipitates.
- *gas injectors* injecting gas into the reservoir often as a means of disposal or sequestering for later production, but also to maintain reservoir pressure.

Lahee Classification:

- *New Field Wildcat* (NFW) – far from other producing fields and on a structure that has not previously produced.
- *New Pool Wildcat* (NPW) – new pools on already producing structure.
- *Deeper Pool Test* (DPT) – on already producing structure and pool, but on a deeper pay zone.
- *Shallower Pool Test* (SPT) – on already producing structure and pool, but on a shallower pay zone.
- *Outpost* (OUT) – usually two or more locations from nearest productive area.
- *Development Well* (DEV) – can be on the extension of a pay zone, or between existing wells (*Infill*).

Cost

The cost of a well depends mainly on the daily rate of the drilling rig, the extra services required to drill the well, the duration of the well program (including downtime and weather time), and the remoteness of the location (logistic supply costs). The daily rates of offshore drilling rigs vary by their capability, and the market availability. Rig rates reported by industry web service show that the deepwater water floating drilling rigs are over twice that of the shallow water fleet, and rates for jackup fleet can vary by factor of 3 depending upon capability. With deepwater drilling rig rates in 2010 of around $420,000/day, and similar additional spread costs, a deep water well of duration of 100 days can cost around US$100 million. With high performance jackup rig rates in 2010 of around $150,000, and similar service costs, a high pressure, high temperature well of duration 100 days can cost about US$30 million.

Onshore wells can be considerably cheaper, particularly if the field is at a shallow depth, where costs range from less than $1 million to $15 million for deep and difficult wells.

Drilling Fluid

In geotechnical engineering, drilling fluid is a fluid used to aid the drilling of boreholes into the earth. Often used while drilling oil and natural gas wells and on exploration drilling rigs, drilling fluids are also used for much simpler boreholes, such as water wells. Liquid drilling fluid is often called drilling mud. The three main categories of drilling fluids are water-based muds (which can be dispersed and non-dispersed), non-aqueous muds, usually called oil-based mud, and gaseous drilling fluid, in which a wide range of gases can be used.

The main functions of drilling fluids include providing hydrostatic pressure to prevent formation fluids from entering into the well bore, keeping the drill bit cool and clean during drilling, carrying out drill cuttings, and suspending the drill cuttings while drilling is paused and when the drilling assembly is brought in and out of the hole. The drilling fluid used for a particular job is selected to avoid formation damage and to limit corrosion.

Types of Drilling Fluid

Many types of drilling fluids are used on a day-to-day basis. Some wells require that different types be used at different parts in the hole, or that some types be used in combination with others. The various types of fluid generally fall into a few broad categories:

- Air: Compressed air is pumped either down the bore hole's annular space or down the drill string itself.
- Air/water: The same as above, with water added to increase viscosity, flush the hole, provide more cooling, and/or to control dust.
- Air/polymer: A specially formulated chemical, most often referred to as a type of polymer, is added to the water & air mixture to create specific conditions. A foaming agent is a good example of a polymer.
- Water: Water by itself is sometimes used.
- Water-based mud (WBM): A most basic water-based mud system begins with water, then clays and other chemicals are incorporated into the water to create a homogenous blend resembling something between chocolate milk and a malt (depending on viscosity). The clay (called "shale" in its rock form) is usually a combination of native clays that are suspended in the fluid while drilling, or specific types of clay that are processed and sold as additives for the WBM system. The most common of these is bentonite, frequently referred to in the oilfield as "gel". Gel likely makes reference to the fact that while the fluid is being pumped, it can be very thin and free-flowing (like chocolate milk), though when pumping is stopped, the static fluid builds a "gel" structure that resists flow. When an adequate pumping force is applied to "break the gel", flow resumes and the fluid returns to its previously free-flowing state. Many other chemicals (e.g. potassium formate) are added to a WBM system to achieve various effects, including: viscosity control, shale stability, enhance drilling rate of penetration, cooling and lubricating of equipment.
- Oil-based mud (OBM): Oil-based mud can be a mud where the base fluid is a petroleum product such as diesel fuel. Oil-based muds are used for many reasons, some being increased lubricity, enhanced shale inhibition, and greater cleaning abilities with less viscosity. Oil-based muds also withstand greater heat without breaking down. The use of oil-based muds has special considerations. These include cost and environmental considerations.
- Synthetic-based fluid (SBM) (Otherwise known as Low Toxicity Oil Based Mud or LTOBM): Synthetic-based fluid is a mud where the base fluid is a synthetic oil. This is most often used on offshore rigs because it has the properties of an oil-based

mud, but the toxicity of the fluid fumes are much less than an oil-based fluid. This is important when men work with the fluid in an enclosed space such as an offshore drilling rig.

On a drilling rig, mud is pumped from the *mud pits* through the drill string where it sprays out of nozzles on the drill bit, cleaning and cooling the drill bit in the process. The mud then carries the crushed or cut rock ("cuttings") up the annular space ("annulus") between the drill string and the sides of the hole being drilled, up through the surface *casing,* where it emerges back at the surface. Cuttings are then filtered out with either a [shale shaker], or the newer shale conveyor technology, and the mud returns to the *mud pits.* The mud pits let the drilled "fines" settle; the pits are also where the fluid is treated by adding chemicals and other substances.

The returning mud can contain natural gases or other flammable materials which will collect in and around the shale shaker / conveyor area or in other work areas. Because of the risk of a fire or an explosion if they ignite, special monitoring sensors and explosion-proof certified equipment is commonly installed, and workers are advised to take safety precautions. The mud is then pumped back down the hole and further re-circulated. After testing, the mud is treated periodically in the mud pits to ensure properties which optimize and improve drilling efficiency, borehole stability, and other requirements listed below.

Function

The main functions of a *drilling mud* can be summarised as follows:

Remove Cuttings from Well

Drilling fluid carries the rock excavated by the drill bit up to the surface. Its ability to do so depends on cutting size, shape, and density, and speed of fluid travelling up the well (annular velocity). These considerations are analogous to the ability of a stream to carry sediment; large sand grains in a slow-moving stream settle to the stream bed, while small sand grains in a fast-moving stream are carried along with the water. The mud viscosity is another important property, as cuttings will settle to the bottom of the well if the viscosity is too low. Other properties include:

- Most drilling muds are thixotropic (that is, they become a gel under static conditions). This characteristic keeps the cuttings suspended when the mud is not moving during, for example, maintenance.
- Fluids that have shear thinning and elevated viscosities are efficient for hole cleaning.

- Higher annular velocity improves cutting transport. Transport ratio (transport velocity / lowest annular velocity) should be at least 50%.
- High density fluids may clean hole adequately even with lower annular velocities (by increasing the buoyancy force acting on cuttings). But may have a negative impact if mud weight is in excess of that needed to balance the pressure of surrounding rock (formation pressure), so mud weight is not usually increased for hole cleaning purposes.
- Higher rotary drill-string speeds introduce a circular component to annular flow path. This helical flow around the drill-string causes drill cuttings near the wall, where poor hole cleaning conditions occur, to move into higher transport regions of the annulus.

Suspend and Release Cuttings

- Must suspend drill cuttings, weight materials and additives under a wide range of conditions.
- Drill cuttings that settle can causes bridges and fill, which can cause stuck-pipe and lost circulation.
- Weight material that settles is referred to as sag, this causes a wide variation in the density of well fluid, this more frequently occurs in high angle and hot wells
- High concentrations of drill solids are detrimental to:
  - o Drilling efficiency (it causes increased mud weight and viscosity, which in turn increases maintenance costs and increased dilution)
  - o Rate of Penetration (ROP) (increases horsepower required to circulate)
  - o Mud properties that suspended must balanced with properties in cutting removal by solids control equipment
- For effective solids controls, drill solids must be removed from mud on the 1st circulation from the well. If re-circulated, cuttings break into smaller pieces and are more difficult to remove.
- Conduct a test to compare the sand content of mud at flow line and suction pit (to determine whether cuttings are being removed).

Control Formation Pressures

- If formation pressure increases, mud density should also be increased, often with barite (or other weighting materials) to balance pressure and keep the wellbore stable. Unbalanced formation pressures will cause an unexpected influx of pressure in the wellbore possibly leading to a blowout from pressured formation fluids.

- Hydrostatic pressure = density of drilling fluid * true vertical depth * acceleration of gravity. If hydrostatic pressure is greater than or equal to formation pressure, formation fluid will not flow into the wellbore.
- Well control means no uncontrollable flow of formation fluids into the wellbore.
- Hydrostatic pressure also controls the stresses caused by tectonic forces, these may make wellbores unstable even when formation fluid pressure is balanced.
- If formation pressure is subnormal, air, gas, mist, stiff foam, or low density mud (oil base) can be used.
- In practice, mud density should be limited to the minimum necessary for well control and wellbore stability. If too great it may fracture the formation.

Seal Permeable Formations

- When mud column pressure exceeds formation pressure, mud filtrate invades the formation, and a filter cake of mud is deposited on the wellbore wall.
- Mud is designed to deposit thin, low permeability filter cake to limit the invasion.
- Problems occur if a thick filter cake is formed; tight hole conditions, poor log quality, stuck pipe, lost circulation and formation damage.
- In highly permeable formations with large pore throats, whole mud may invade the formation, depending on mud solids size;
    - o Use bridging agents to block large opening, then mud solids can form seal.
    - o For effectiveness, bridging agents must be over the half size of pore spaces / fractures.
- Depending on the mud system in use, a number of additives can improve the filter cake (e.g. bentonite, natural & synthetic polymer, asphalt and gilsonite).

Maintain Wellbore Stability

- Chemical composition and mud properties must combine to provide a stable wellbore. Weight of the mud must be within the necessary range to balance the mechanical forces.

- Wellbore instability = sloughing formations, which can cause tight hole conditions, bridges and fill on trips (same symptoms indicate hole cleaning problems).
- Wellbore stability = hole maintains size and cylindrical shape.
- If the hole is enlarged, it becomes weak and difficult to stabilise, resulting in problems such as low annular velocities, poor hole cleaning, solids loading and poor formation evaluation
- In sand and sandstones formations, hole enlargement can be accomplished by mechanical actions (hydraulic forces & nozzles velocities). Formation damage is reduced by conservative hydraulics system. A good quality filter cake containing bentonite is known to limit bore hole enlargement.
- In shales, mud weight is usually sufficient to balance formation stress, as these wells are usually stable. With water base mud, chemical differences can cause interactions between mud & shale that lead to softening of the native rock. Highly fractured, dry, brittle shales can be extremely unstable (leading to mechanical problems).
- Various chemical inhibitors can control mud / shale interactions (calcium, potassium, salt, polymers, asphalt, glycols and oil – best for water sensitive formations)
- Oil (and synthetic oil) based drilling fluids are used to drill most water sensitive Shales in areas with difficult drilling conditions.
- To add inhibition, emulsified brine phase (calcium chloride) drilling fluids are used to reduce water activity and creates osmotic forces to prevent adsorption of water by Shales.

Minimising Formation Damage

- Skin damage or any reduction in natural formation porosity and permeability (washout) constitutes formation damage
- Most common damage;
  - o Mud or drill solids invade the formation matrix, reducing porosity and causing skin effect
  - o Swelling of formation clays within the reservoir, reduced permeability
  - o Precipitation of solids due to mixing of mud filtrate and formations fluids resulting in the precipitation of insoluble salts
  - o Mud filtrate and formation fluids form an emulsion, reducing reservoir porosity
- Specially designed drill-in fluids or workover and completion fluids, minimise formation damage.

Cool, Lubricate, and Support the Bit and Drilling Assembly

- Heat is generated from mechanical and hydraulic forces at the bit and when the drill string rotates and rubs against casing and wellbore.
- Cool and transfer heat away from source and lower to temperature than bottom hole.
- If not, the bit, drill string and mud motors would fail more rapidly.
- Lubrication based on the coefficient of friction. Oil- and synthetic-based mud generally lubricate better than water-based mud (but the latter can be improved by the addition of lubricants).
- Amount of lubrication provided by drilling fluid depends on type & quantity of drill solids and weight materials + chemical composition of system.
- Poor lubrication causes high torque and drag, heat checking of the drill string, but these problems are also caused by key seating, poor hole cleaning and incorrect bottom hole assemblies design.
- Drilling fluids also support portion of drill-string or casing through buoyancy. Suspend in drilling fluid, buoyed by force equal to weight (or density) of mud, so reducing hook load at derrick.
- Weight that derrick can support limited by mechanical capacity, increase depth so weight of drill-string and casing increase.
- When running long, heavy string or casing, buoyancy possible to run casing strings whose weight exceed a rig's hook load capacity.

Transmit Hydraulic Energy to Tools and Bit

- Hydraulic energy provides power to mud motor for bit rotation and for MWD (measurement while drilling) and LWD (logging while drilling) tools. Hydraulic programs base on bit nozzles sizing for available mud pump horsepower to optimize jet impact at bottom well.
- Limited to:
  - Pump horsepower
  - Pressure loss inside drillstring
  - Maximum allowable surface pressure
  - Optimum flow rate

  - o Drill string pressure loses higher in fluids higher densities, plastic viscosities and solids.
- Low solids, shear thinning drilling fluids such as polymer fluids, more efficient in transmit hydraulic energy.
- Depth can be extended by controlling mud properties.
- Transfer information from MWD & LWD to surface by pressure pulse.

Ensure Adequate Formation Evaluation

- Chemical and physical mud properties and wellbore conditions after drilling affect formation evaluation.
- Mud loggers examine cuttings for mineral composition, visual sign of hydrocarbons and recorded mud logs of lithology, ROP, gas detection or geological parameters.
- Wireline logging measure – electrical, sonic, nuclear and magnetic resonance.
- Potential productive zone are isolated and performed formation testing and drill stem testing.
- Mud helps not to disperse of cuttings and also improve cutting transport for mud loggers determine the depth of the cuttings originated.
- Oil-based mud, lubricants, asphalts will mask hydrocarbon indications.
- So mud for drilling core selected base on type of evaluation to be performed (many coring operations specify a blend mud with minimum of additives).

Control Corrosion (in Acceptable Level)

- Drill-string and casing in continuous contact with drilling fluid may cause a form of corrosion.
- Dissolved gases (oxygen, carbon dioxide, hydrogen sulfide) cause serious corrosion problems;
  - o Cause rapid, catastrophic failure
  - o May be deadly to humans after a short period of time
- Low pH (acidic) aggravates corrosion, so use corrosion coupons to monitor corrosion type, rates and to tell correct chemical inhibitor is used in correct amount.
- Mud aeration, foaming and other O_2 trapped conditions cause corrosion damage in short period time.

- When drilling in high H_2S, elevated the pH fluids + sulfide scavenging chemical (zinc).

Facilitate Cementing and Completion

- Cementing is critical to effective zone and well completion.
- During casing run, mud must remain fluid and minimise pressure surges so fracture induced lost circulation does not occur.
- Mud should have thin, slick filter cake, wellbore with no cuttings, cavings or bridges.
- To cement and completion operation properly, mud displace by flushes and cement. For effectiveness;
  - o Hole near gauges
  - o Mud low viscosity
  - o Mud non progressive gel strength.

Minimise Impact on Environment

Mud is, in varying degrees, toxic. It is also difficult and expensive to dispose of it in an environmentally friendly manner. A Vanity Fair article described the conditions at Lago Agrio, a large oil field in Ecuador where drillers were effectively unregulated.

Composition of Drilling Mud

Water-based drilling mud most commonly consists of bentonite clay (gel) with additives such as barium sulfate (barite), calcium carbonate (chalk) or hematite. Various thickeners are used to influence the viscosity of the fluid, e.g. xanthan gum, guar gum, glycol, carboxymethylcellulose, polyanionic cellulose (PAC), or starch. In turn, deflocculants are used to reduce viscosity of clay-based muds; anionic polyelectrolytes (e.g. acrylates, polyphosphates, lignosulfonates (Lig) or tannic acid derivates such as Quebracho) are frequently used. Red mud was the name for a Quebracho-based mixture, named after the colour of the red tannic acid salts; it was commonly used in 1940s to 1950s, then was made obsolete when lignosulfonates became available. Other components are added to provide various specific functional characteristics as listed above. Some other common additives include lubricants, shale inhibitors, fluid loss additives (to control loss of drilling fluids into permeable formations). A weighting agent such as barite is added to increase the overall density of the drilling fluid so that sufficient bottom hole pressure can be maintained thereby preventing an unwanted (and often dangerous) influx of formation fluids.

Mud Engineer

"Mud engineer" is the name given to an oil field service company individual who is charged with maintaining a drilling fluid or completion fluid system on an oil and/or gas drilling rig. This individual typically works for the company selling the chemicals for the job and is specifically trained with those products, though independent mud engineers are still common. The work schedule of the mud engineer or more properly Drilling Fluids Engineer, is arduous, often involving long shifts. Typical shifts range from twenty-one days on and seven off, to five weeks on and five weeks off. Consulting mud men can work up to one hundred days or longer, in a row. Nowadays, the internationally accepted shift pattern at off-shore drilling operations is personnel (including mud engineers) work on a 28 day shift pattern, where they work for 28 continuous days and rest the following 28 days. In Europe this is more commonly a 21 day shift pattern.

In offshore drilling, with new technology and high total day costs, wells are being drilled extremely fast. Having two mud engineers makes economic sense to prevent down time due to drilling fluid difficulties. Two mud engineers also reduce insurance costs to oil companies for environmental damage that oil companies are responsible for during drilling and production. A senior mud engineer typically works in the day, and a junior mud engineer at night.

The cost of the drilling fluid is typically about 10% (may vary greatly) of the total cost of drilling a well, and demands competent mud engineers. Large cost savings result when the mud engineer and fluid performs adequately.

The mud engineer is not to be confused with mudloggers, service personnel who monitor gas from the mud and collect well bore samples.

Compliance Engineer

The compliance engineer is the most common name for a relatively new position in the oil field, emerging around 2002 due to new environmental regulations on synthetic mud in the United States. Previously, synthetic mud was regulated the same as water-based mud and could be disposed of in offshore waters due to low toxicity to marine organisms. New regulations restrict the amount of synthetic oil that can be discharged. These new regulations created a significant burden in the form of tests needed to determine the "ROC" or retention on cuttings, sampling to determine the percentage of crude oil in the drilling mud, and extensive documentation. It should be noted that no type of oil/synthetic based mud (or drilled cuttings contaminated with

OBM/SBM) may be dumped in the North Sea. Contaminated mud must either be shipped back to shore in skips or processed on the rigs.

A new monthly toxicity test is also now performed to determine sediment toxicity, using the amphipod *Leptocheirus plumulosus*. Various concentrations of the drilling mud are added to the environment of captive *L. plumulosus* to determine its effect on the animals. The test is controversial for two reasons:

1. These animals are not native to many of the areas regulated by them, including the Gulf of Mexico
2. The test has a very large standard deviation and samples that fail badly may pass easily upon retesting.

Chapter 3

Pumping

In the context of oil wells, pumping is a routine operation involving injecting fluids into the well. Pumping may either be done by rigging up to the kill wing valve on the Xmas tree or, if an intervention rig up is present pumping into the riser through a T-piece (a small section of riser with a connection on the side). Pumping is most routinely done to protect the well against scale and hydrates through the pumping of scale inhibitors and methanol. Pumping of kill weight brine may be done for the purposes of well kills and more exotic chemicals may be pumped from surface for cleaning the lower completion or stimulating the reservoir (though these types are jobs are more frequently done with coiled tubing for extra precision).

Importance of Knowing Quantity

Work involving wells is fraught with difficulties as there is often very little information about the real time condition of the completion. This lack of knowledge also covers potential damage and even loss of well integrity. Therefore it is essential for the operator to pay attention to the pressures as recorded and to the quantity pumped. A premature increase in pressure is sign of a potential blockage and continuing to pump risks burst pressure retaining components. Pumping more than an anticipated amount of fluid is a sign of a loss of integrity and a potential leak path somewhere. In either of these two situations, pumping must be stopped and the potential causes analysed.

Compressed Volumes

It is vital to know the effective capacity of the completion being filled in order to understand what are sensible volumes. If pumping is to continue until reaching a desired pressurisation, then the compressibility of the fluid will become significant. It is therefore important to know how much the fluid will compress under pressure to know how much extra fluid is expected to be required.

As a rule of thumb in the oilfield, compression is governed by the equation:

$\Delta V = P \times V \times k$ where ΔV is the change in volume, P is the pressure at surface and V is the volume of fluid unpressurised. k is a compression factor approximately 3.5×10^{-6} psi^{-1}.

For example, a volume of 300 bbl is to be filled with brine and pressurised to 3000 psi at the surface. The compression is

$$\Delta V = PVk$$

$$\Delta V = 3000 psi \times 300 bbl \times 3.5 \times 10^{-6} psi^{-1}$$

$$\Delta V = 3.15 bbl$$

Therefore, it is expected that 303.15 bbl are required to accomplish this task. If 3000 psi is achieved prior to this quantity being pumped, a blockage is to be suspected. If after pumping 303 bbl, pressurisation is not achieved, a leak is to be suspected.

Well Integrity

Well integrity, in regards to oil wells, is defined by NORSOK D-010 as the “”Application of technical, operational and organisational solutions to reduce risk of uncontrolled release of formation fluids throughout the life cycle of a well”. There are various facets to well integrity, including accountability/responsibility, well operating processes, well service processes, tubing/annulus integrity, tree/ wellhead integrity and testing of safety systems.

Accountability/Responsibility

Accountability is the position that ensures a particular task is being done. They typically control the budget and organisational chart structure. Responsibility is the position that actually does the task. There are a variety of tasks associated with well integrity and responsibility can reside with a field’s well operations engineer, operators, well service technicians, etc.

Well Operating Processes

This includes processes such as personnel competency, well startup-operating-shutdown procedures, process to report anomalies, corrosion/erosion control, etc.

Well Service Processes

This includes processes such as personnel competency, intervention procedures, etc.

Tubing/Annulus Integrity

This refers to the integrity of the production tubing. Common threats to tubing integrity are erosion and corrosion by production fluids, which denude the steel. The other threat is pressure differential between the tubing and the 'A' annulus. If it exceeds the rated pressure of the tubing, there is a risk of tubing burst or tubing collapse.

This refers to the integrity of the casing strings, which bound the annuli. All annuli will naturally be filled to some level with brine or other fluids from the completion, though the 'A' annulus may have gas down to the lowest gas lift valve if the well is gas lifted. The major integrity issues comes from the differential pressure across the casings due to the weight of brine on either side. To protect well integrity, each annulus will have a Maximum Allowable Annulus Surface Pressure.

Tree and Wellhead Integrity

This refers to the integrity of the surface (or subsea) equipment. The wellhead and the tree are typically suitably engineered to be able to withstand the normal operating pressures. The major integrity issues for these components surround the operation of the valves, which are prone to leaking. For this reason, valves must be routinely maintained.

Wireline (Cabling)

In the oil and gas industry, the term wireline usually refers to a cabling technology used by operators of oil and gas wells to lower equipment or measurement devices into the well for the purposes of well intervention and reservoir evaluation.

Braided line can contain an inner core of insulated wires which provide power to equipment located at the end of the cable, normally referred to as electric line, and provides a pathway for electrical telemetry for communication between the surface and equipment at the end of the cable.

Wireline Tools

A wireline tool string can be dozens of feet long with multiple separate tools installed to perform multiple operations at once.

Natural gamma-ray tools are designed to measure naturally occurring gamma radiation in the earth caused by the disintegration due to Potassium, Uranium, and Thorium. Unlike nuclear tools, these natural gamma ray tools do not emit any radiation.

Natural gamma ray tools employ a radioactive sensor, which is usually a scintillation crystal that emits a light pulse proportional to the strength of the gamma ray pulse incident on it. This light pulse is then converted to a current pulse by means of a photo multiplier tube PMT. From the photo multiplier tube, the current pulse goes to the tool's electronics for further processing and ultimately to the surface system for recording. The strength of the received gamma rays is dependent on the source emitting gamma rays, the density of the formation, and the distance between the source and the tool detector. The log recorded by this tool is used to identify lithology, estimate shale content, and depth correlation of future logs.

Nuclear Tools

Nuclear tools measure formation properties through the interaction of reservoir molecules with radiation emitted from the logging tool. Most open hole nuclear tools utilise double-encapsulated chemical sources.

Density Tools

Density tools use gamma ray radiation to determine the lithology and porosity of the well environment. Modern density tools utilise a Cs-137 radioactive source to generate gamma rays. Density tools also have an extendable caliper arm, which is used to measure the true width of the borehole.

Gamma rays emitted from the source pass into the formation. Depending on the density of the surrounding formation, some of the gamma rays will be absorbed into the rock while others are reflected back to the tool. The ratio of returning gamma rays to absorbed gamma rays is useful in determining formation density.

Neutron Tools

Neutron tools use fast neutrons to indicate porosity and lithology of the well. Neutron tools typically contain an Am241-Be chemical source or Minitron electronic source to generate the neutrons.

The hydrogen content of the formation, from oil or water, slows down the emitted neutrons until they reach a thermal or epithermal state. At the slower thermal and epithermal states, the tool is able to detect the neutrons. These counts therefore yield a count of slow neutrons, which is a clear indicator of the hydrogen content of the well.

Resistivity Tools

This tool is important in reservoir evaluation for determining the location of the oil-water contact. Water is far more conductive than hydrocarbons and so will give the reservoir rock it saturates a lower resistivity than rock saturated with hydrocarbons. When analysing a resistivity log, the point where the resistivity undergoes a large change is likely to be the location of the oil-water contact. It is also used as an indicator for permeability. Since most resistivity tools have different depths of investigation, a permeable formation will read different resistivities at different depths.

Sonic and Ultrasonic Tools

Sonic tools generate sound wave and measure the time it takes to reach the detectors. This is used to measure the effective porosity. Sound waves travel slower in formations in which the pores are not interconnected.

Magnetic Resonance Tools

A measurement of the nuclear magnetic resonance (NMR) properties of hydrogen in the formation. There are two phases to the measurement: polarization and acquisition. First, the hydrogen atoms are aligned in the direction of a static magnetic field (B0). This polarization takes a characteristic time T1. Second, the hydrogen atoms are tipped by a short burst from an oscillating magnetic field that is designed so that they precess in resonance in a plane perpendicular to B0. The frequency of oscillation is the Larmor frequency. The precession of the hydrogen atoms induces a signal in the antenna. The decay of this signal with time is caused by transverse relaxation and is measured by the CPMG pulse sequence. The decay is the sum of different decay times, called T2. The T2 distribution is the basic output of a NMR measurement.

The NMR measurement made by both a laboratory instrument and a logging tool follow the same principles very closely. An important feature of the NMR measurement is the time needed to acquire it. In the laboratory, time presents no difficulty. In a log, there is a trade-off between the time needed for polarization and acquisition, logging speed and frequency of sampling. The longer the polarization and acquisition, the more complete the measurement. However, the longer times require either lower logging speed or less frequent samples.

Cement Bond Tools

A cement bond tool, or CBT, is an acoustic tool used to measure the quality of the cement behind the casing. Using a CBT, the bond between the casing and cement as well as the bond between cement and formation can be determined. Using CBT data, a company can troubleshoot problems with the cement sheath if necessary. This tool must be centralised in the well to function properly.

Two of the largest problems found in cement by CBT's are channelling and micro-annulus. A micro annulus is the formation of microscopic cracks in the cement sheath. Channelling is where large, contiguous voids in the cement sheath form, typically caused by poor centralisation of the casing. Both of these situations can, if necessary, be fixed by remedial electric line work.

A CBT gains its measurements by rapidly pulsing out compressional waves across the well bore and into the pipe, cement, and formation. The compressional pulse originates in a transmitter at the top of the tool, which, when powered up on surface sounds like a rapid clicking sound. The tool typically has two receivers, one three feet away from the receiver, and another at five feet from the transmitter. These receivers record the arrival time of the compressional waves. The information from these receivers are logged as traveltimes for the three and five foot receivers and as a micro-seismogram.

Recent advances in logging technologies have allowed the receivers to measure 360 degrees of cement integrity and can be represented on a log as a radial cement map and as 6-8 individual sector arrival times.

Casing Collar Locators

Casing collar locator tools, or CCL's, are among the simplest and most essential in cased hole electric line. CCL's are typically used for depth correlation and can be an indicator of line overspeed when logging in heavy fluids.

A CCL operates on Faraday's Law of Induction. Two magnets are separated by a coil of copper wire. As the CCL passes by a casing joint, or collar, the difference in metal thickness across the two magnets induces a current spike in the coil. This current spike is sent uphole and logged as what's called a collar kick on the cased hole log.

Gamma Perforating Tools

A cased hole gamma perforator is used to perform mechanical services, such as shooting perforations, setting downhole tubing/casing elements, dumping remedial cement, tracer surveys, etc. Typically, a gamma perforator will have some sort of explosively initiated device attached to it, such as a perforating gun, a setting tool, or a dump bailor. In certain instances, the gamma perforator is used to merely spot objects in the well, as in tubing conveyed perforating operations and tracer surveys. Gamma perforators operate in much the same way as an open hole natural gamma ray tool. Gamma rays given off from naturally occurring radioactive elements bombard the tool. The tool processes the gamma ray counts and sends the data uphole where it is put onto a log. The information is then used to ensure that the depth shown on the log is correct. After that, power can be applied through the tool to set off explosive charges for things like perforating, setting plugs or packers, dumping cement, etc.

Setting Tools

Setting tools are used to set downhole completion elements. Setting tools are typically large steel tools onto which a downhole completion can be screwed. One of the most common setting tools is manufactured by Baker Hughes. Setting tools are explosively driven devices. A shooting CCL or a gamma perforator is used to apply power to detonate a low explosive in the setting tool. The gas pressure created by the deflagrating low explosive exerts a large force on a piston holding back oil. The pneumatic pressure of the piston pushes the oil, which hydraulically separates the setting tool from the plug or packer. The downhole completion is now set in place. Not only for completions, Setting tools can also run bridge plugs. Which are most commonly used to abandon a well. A certain amount of oil well cement must then be placed on top of the plug. A bond log is also common protocol, the cement must be bonded with the casing to abandon a well, if not, there must be squeeze guns shot.

Additional Equipment

Cable Head

The cable head is the upper most portion of the toolstring on any given type of wireline. The cable head is where the conductor wire is made into an electrical connection that can be connected to the rest of the toolstring. Cable heads are typically custom built by the wireline operator for every job and depend greatly on depth, pressure and the type of wellbore fluid. Electric line weakpoints are also located in the cable head. If the tool is to become stuck in the well, the weak point is where the tool would first separate from the wireline. If the wireline were severed anywhere else along the line, the tool becomes much more difficult to fish.

Tractors

These are electrical tools used to push the toolstring into hole, overcoming wireline's disadvantage of being gravity dependent. These are used for in highly deviated and horizontal wells where gravity is insufficient, even with roller stem. They push against the side of the wellbore either through the use of wheels or through a wormlike motion. The technology has been in place for more than 10 years, Welltec has operation factors of over 98% with their wireline tractors. The leading operator on the Norwegian Continental Shelf, has successfully applied this technology since 1996 and has concluded that it is a reliable as well as a cost-efficient technology. According to the group's calculations, they save approximately NOK 500 million ($80 million USD) annually on tractor operations and from 1996 to 2005, tractors have covered an accumulated distance of more than 3,000 kilometres through horizontal wells for the company.

Measuring Head

A measuring head is the first piece of equipment the wireline comes into contact with off the drum. The measuring head is composed of several wheels which support the wireline on its way to the winch and they also measure crucial wireline data. A measuring head records tension, depth, and speed. Current models use optical encoders to derive the revolutions of a wheel with a known circumference, which in turn is used to figure speed and depth. A wheel with a pressure sensor is used to figure tension.

Wireline Apparatus

For oilfield work, the wireline resides on the surface, wound around a large (3 to 10 feet in diametre) spool. Operators may use a portable spool (on the back of a special truck) or a permanent part of the drilling rig. A motor and drive train turn the spool and raise and lower the equipment into and out of the well – the winch.

Pressure Control During Wireline Operations

The pressure control employed during wireline operations is intended to contain pressure originating from the well bore. During open hole electric line operations, the pressure might be the result from a well kicking. During cased hole electric line, this is most likely the result of a well producing at high pressures. Pressure equipment must be rated to well over the expected well pressures.

Flange

A flange attaches to the top of the Christmas tree, usually with some sort of adapter for the rest of the pressure control. A metal gasket is placed between the top of the Christmas tree and the flange to keep in well pressures.

Wireline Valve

A wireline valve, also called a wireline blow out preventer(BOP), is an enclosed device with one or more rams capable of closing over the wireline in an emergency. A dual wireline valve has two sets of rams and some have the capability of pumping grease in the space between the rams to counterbalance the well pressure.

Lubricator

Lubricator is the term used for sections of pressure tested pipe that act to seal in wireline tools during pressurization.

Pump-In Sub

Pump-in subs allow for the injection of fluid into the pressure control string. Normally these are used for wellsite pressure testing, which is typically performed between every run into the well. They can also be used to bleed off pressure from the string after a run in the well, or to pump in kill fluids to control a wild well.

Grease Injector Head

The grease injector head is the main apparatus for controlling well pressure while running into the hole. The grease head uses a series of very small pipes, called flow tubes, to decrease the pressure head of the well. Grease is injected at high pressure into the bottom portion of the grease head to counteract the remaining well pressure.

Pack-Off Sub

Pack-off subs utilise grease pressure on a rubber sealing element to create an impermeable seal around the wireline. Pack-off subs can be hand pumped or compressed through a motorized pumping unit.

Line Wiper

A line wiper operates in much the same way as a pack-off sub, except that the rubber element is much softer. Grease pumps exert force on the rubber element until a light pressure is exerted on the wireline, cleaning grease and well fluid off the line in the process.

Quick Test Sub

A Quick Test Sub (QTS) is used when pressure testing the pressure control equipment (PCE) that will be used during explosives operations. The PCE is pressure tested and then broke at the QTS. The explosives are then attached to the tool string and pulled back in to the lubricator. The PCE is then reconnected at the QTS. The QTS has two O-rings where it was disconnected that can be tested with hydraulic pressure to confirm the PCE can still hold the pressure it was tested to.

Ball-Check Valve

If the wireline were to become severed from the tool, a ball check valve can seal the well off from the surface. During wireline operations, a steel ball sits to the side of a confined area within the grease head while the cable runs in and out of the hole. If the wireline exits that confined area under pressure, the pressure will force the steel ball up towards the hole where the wireline had been. The ball's diametre is larger than that of the hole, so the ball effectively seals off pressure to the surface.

Head Catcher

A head catcher is a device placed at the top of the lubricator section. Should the wireline tools be forced into the top of the lubricator section, the head catcher, which looks like a small 'claw,' will clamp down on the fishing neck of the tool. This action prevents the tools from falling downhole should the line pull out of the rope socket. Pressure is bled off of the head catcher to release the tools.

Tool Trap

A tool trap has the same purpose as a head catcher in that it prevents the tools from inadvertently dropping down the hole.

Coiled Tubing

In the oil and gas industries, coiled tubing refers to metal piping, normally 1" to 3.25" in diametre, used for interventions in oil and gas wells and sometimes as production tubing in depleted gas wells, which comes spooled on a large reel. Coiled tubing is often used to carry out operations similar to wirelining. The main benefits over wireline are the ability to pump chemicals through the coil and the ability to push it into the hole rather than relying on gravity. However, for offshore operations, the 'footprint' for a coiled tubing operation is generally larger than a wireline spread, which can limit the number of installations where coiled tubing can be performed and make the operation more costly. A coiled tubing operation is normally performed through the drilling derrick on the oil platform, which is used to support the surface equipment, although on platforms with no drilling facilities a self supporting tower can be used instead. For coiled tubing operations on sub-sea wells a Mobile Offshore Drilling Unit (MODU) e.g. semi-submersible, Drillship etc. has to be utilised to support all the surface equipment and personnel, whereas wireline can be carried out from a smaller and cheaper intervention vessel. Onshore, they can be run using smaller service rigs, and for light operations a mobile self-contained coiled tubing rig can be used.

The tool string at the bottom of the coil is often called the bottom hole assembly (BHA). It can range from something as simple as a jetting nozzle, for jobs involving pumping chemicals or cement through the coil, to a larger string of logging tools, depending on the operations.

Coil tubing has also been used as a cheaper version of work-over operations. It is used to perform open hole drilling and milling operations. It can also be used to fracture the reservoir, a process where fluid is pressurised to thousands of psi on a specific point in a well to literally break the rock apart and allow the flow of product. Coil tubing can perform almost any operation for oil well operations if used correctly.

Uses

Circulation: The most popular use for coiled tubing is circulation or deliquification. A hydrostatic head (a column of fluid in the well bore) may be inhibiting flow of formation fluids due to its weight (the well is said to have been killed). The safest (though not the cheapest) solution would be to attempt to circulate out the fluid, using a gas, frequently nitrogen (*Often called a 'Nitrogen Kick'*). By running coiled tubing into the bottom of the hole and pumping in the gas, the kill fluid can be forced out to production. Circulating can also be used to clean out light debris, which may have accumulated in the hole. Coiled tubing umbilicals can convey hydraulic submersible pumps and jet pumps into wells. These pumps allow for inexpensive and non invasive well cleanouts on low pressure CBM (coal bed methane) gas wells. These umbilicals can also be run into deviated wells and horizontal laterals.

Pumping: Pumping through coiled tubing can also be used for dispersing fluids to a specific location in the well such as for cementing perforations or performing chemical washes of downhole components such as sandscreens. In the former case, coiled tubing is particularly advantageous compared to simply pumping the cement from surface as allowing it to flow through the entire completion could potentially damage important components, such as the downhole safety valve. Coiled tubing umbilical technologies enable the deployment of complex pumps which require multiple fluid strings on coiled tubing. In many cases, the use of coiled tubing to deploy a complex pump can greatly reduce the cost of deployment by eliminating the number of units on site during the deploy.

Drilling: A relatively modern drilling technique involves using coiled tubing instead of conventional drill pipe. This has the advantage of requiring less effort to trip in and out of the well (the coil can simply be run in and pulled out while drill string must be assembled and dismantled joint by joint while tripping in and out). Additionally, the coiled tubing is stripped into and out of hole, providing a hermetic seal around the coil and, if desired, allowing the well to flow during drilling operations. Instead of rotating the drill bit by using a rotary table or top drive at the surface, it is turned by a downhole motor, powered by the motion of drilling fluid pumped from surface. Drilling which is powered by a mud motor instead of a rotating pipe is generally called slide drilling. When carrying out directional or horizontal drilling a downhole orienter is an essential part of the bottom hole assembly, so that the tool can be rotated to change the direction.

Logging and Perforating: These tasks are by default the realm of wireline. Because coiled tubing is rigid, it can be pushed into the well from the surface. This is an advantage over wireline, which depends on the weight of the toolstring to be lowered into the well. For highly deviated and horizontal wells, gravity may be insufficient for wireline logging. Roller stem and tractors can often overcome this disadvantage at greatly reduced cost, particularly on small platforms and subsea wells where coiled tubing would require mobilising an expensive mobile drilling rig. The use of coiled tubing for these tasks is usually confined to occasions where it is already on site for another purpose, for example a logging run following a chemical wash.

Production: Coiled tubing is often used as a production string in shallow gas wells that produce some water. The narrow internal diametre results in a much higher velocity than would occur inside conventional tubing or inside the casing. This higher velocity assists in lifting liquids to surface, liquids which might otherwise accumulate in the wellbore and eventually "kill" the well. The coiled tubing may be run inside the casing instead or inside conventional tubing. When coiled tubing is run inside of conventional tubing it is often referred to as a "velocity string" and the space between the outside of the coiled tubing and the inside of the conventional tubing is referred to as the "micro annulus". In some cases gas is produced up into the micro annulus. Coiled tubing umbilicals can convey hydraulic submersible pumps, electric submersible pumps and jet pumps into wells for both permanent deliquification schemes and service applications.

Coiled Tubing Rigup

The main engine of a coiled tubing intervention is the injector head. This component contains the mechanism to push and pull the coil in and out of the hole. An injector head has a curved guide beam on top called a gooseneck which threads the coil into the injector body.

Below the injector is the stripper, which contains rubber pack off elements providing a seal around the tubing to isolate the well's pressure. Below the stripper is the blowout preventer, which provides the ability to cut the coiled tubing pipe and seal the well bore (shear-blind) and hold and seal around the pipe (pipe-slip). Older quad-BOPs have a different ram for each of these functions (blind, shear, pipe, slip). Newer dual-BOPs combine some of these functions together to need just two distinct rams (shear-blind, pipe-slip).

The BOP sits on top of the riser, which provides the pressurised tunnel down to the top of the Xmas tree. Between the Xmas tree and the riser is the final pressure barrier, the shear-seal BOP, which can cut and seal the pipe.

Onshore Light Coiled Tubing Unit

A Coil Tubing Unit is a self contained multi-use machine that can approximately do anything a conventional service rig is capable of - with the exception of tripping jointed pipe. There are generally two types in shallow service - Arch and Picker. One uses a vertical elevator with a horsehead on top, and an injector hanging by winch line off it. The Picker units have a picker, and a horsehead bolted directly to the injector. These type of coil tubing units have a permanent drum mounted amidships (They are generally tandem drive Class 3 trucks, 40 feet (12 m) long or so), and a large air compressor, usually good for 2500 psi at 660 CFM, mounted between the drum and cab. In lower pressure, natural gas wells, with no hydrocarbons, the compressor is actually used to blow air to bottom hole in these live natural gas wells, for the purpose of "cleaning out" mud and fluid from the wellbore and perforations. In higher pressure wells, or oil wells, nitrogen or carbon dioxide is the preferred, and much safer method.

Snubbing

Snubbing is a type of heavy well intervention performed on oil and gas wells. It involves running the BHA on a pipe string using a hydraulic workover rig. Unlike wireline or coiled tubing, the pipe is not spooled off a drum but made up and broken up while running in and pulling out, much like conventional drill pipe. Due to the large rigup, it is only used for the most demanding of operations when lighter intervention techniques do not offer the strength and durability. Unlike conventional drilling and completions operations, snubbing can be performed with the well still under pressure (not killed). When done so, it is called hydraulic workover. It can also be performed without having to remove the Xmas tree from the wellhead.

Rigup

A snubbing rigup is a very tall structure. It consists of a hydraulically powered snubbing unit, which provides the force on a pipe, above a string of multi-layered pressure control components.

At the top of the snubbing unit is the basket, which serves as the control post for the rigup. Below the basket are the hydraulic jacks, which power the pipe into and out of the hole. It consists of two mechanisms for applying force to the pipe in either direction. Each mechanism consists of travelling and stationary slips. The travelling slips are used to move the pipe, while the stationary slips are used to hold the pipe while the travelling slips are repositioned between strokes. Primary pressure control during snubbing operations is provided by a self-energising stripper or through an assembly of stripping rams. The stripping rams consists of two sets of rams, one on top of the other. There are opened one at a time to allow the collar of the connections between joints of pipe to pass through the rams. Below the stripping rams are the safety rams, and below them, a shear/blind blowout preventer and an annular BOP.

Stripping the Pipe

Unlike coiled tubing or wireline, where the wire or tubing is always the same diametre allowing for a single unmoving primary barrier (stuffing box or stripper), snubbing uses a pipe, which will have an enlarged collar at the connection between the joints. Therefore, the pressure control system must be able to accommodate this variable diametre. The stripping rams accomplish this. The first stage of lowering a collar through the stripping system is to close the lower rams so as to seal off the mechanism above from wellbore pressure. The space between the rams can then be bled off allowing the upper rams to be opened. The collar can then pass through the opened upper rams. Once the collar is in between the rams, the upper rams are closed and pressure is equalised either side of the lower rams. The lower rams are then safely opened and the collar is lowered through the rams.

This process is repeated as successive collars are lowered into the well. When pulling out of hole, this procedure is reversed.

Another popular method of stripping tubulars in/out of a wellbore is with the use of an "Annular" Blow Out Preventer (BOP). An Annular BOP consists of a natural or synthetic rubber element with encased metal reinforcement sections. A hydraulic piston pushes the annular element up into a concaved cap which forces the element diametre to decrease in size. When the element diametre is closed sufficiently it forms a seal around the body of the pipe. The upset or larger diametre section of a pipe connection can be pulled or pushed through a closed annular element without damage and while still maintaining a gas tight seal. Annular BOPs come in various sizes and pressure ratings and are ideal for lower pressure gas wells. Generally, the maximum pressure for stripping pipe through an annular is equal to 40% of the maximum static pressure rating dry or 60% if the pipe is lubricated as it is being stripped through the annular.

Heavy-Pipe and Light-Pipe

Because snubbing is normally done under pressure, initially, the weight of pipe in wellbore is less than the force due to the wellbore pressure. This is light-pipe. This means that downward force is required on the pipe to force it in against resistance. Once a sufficient amount of pipe has been run in hole, the weight becomes sufficient to overpower the wellbore pressure and the pipe naturally wants to fall in the hole. This is heavy-pipe. At this point, the snubbing mechanism is changed over to the one, which provided upward force to hold the pipe and lower it controllably in the well. When pulling out of hole, upward force is initially used to lift the pipe until the equilibrium point, when the downward force is used to prevent wellbore pressure from blowing the light-pipe out of hole.

Risks

The more complex method of pressure control, as compared to coiled tubing and wireline, naturally invites more opportunity for things to go wrong. One such peril was seen in June 2007 on the Shearwater platform. Snubbing was being used to clean out large pebble, which had entered the well through a collapsed liner. While pulling out of hole, one stripping ram was not opened sufficiently and a collar on the pipe string caught on the ram. The excessive force applied to the pipe caused it to break apart, dropping the string below the failure into the well. In the time it took to prepare to fish out the pipe, the pebbles in the process of being circulated out, settled on the pipe, preventing successful fishing. Although problems such as the one described above can happen they are extremely rare and always avoidable. In the case above, adequate supervision could have prevented this dramatic consequence of operator error by limiting the hydraulic force allowable to be an applied to a level below what was required to part the pipe string.

Workover

The term workover is used to refer to any kind of oil well intervention involving invasive techniques, such as wireline, coiled tubing or snubbing. More specifically though, it will refer to the expensive process of pulling and replacing a completion.

Reason to Perform a Workover

Workovers rank among the most complex, difficult and expensive types of wellwork there is. They are only performed if the completion of a well is terminally unsuitable for the job at hand. The production tubing may have become damaged due to operational factors like corrosion to the point where well integrity is threatened. Downhole components such as tubing retrievable downhole safety valves or electrical submersible pumps may have malfunctioned, needing replacement.

In other circumstances, the reason for a workover may not be that the completion itself is in a bad condition, but that changing reservoir conditions make it unsuitable. For example, a high productivity well may have been completed with 5½" tubing to allow high flow rates (a narrower tubing would have unnecessarily choked the flow). Some years on, declining productivity means the reservoir can no longer support stable flow through this wide bore. This may lead to a workover to replace the 5½" tubing with 4½" tubing.

Operation

Before any workover, the well must first be killed. Since workovers are long planned in advance, there would be much time to plan the well kill and so the reverse circulation would be common. The intense nature of this operation often requires no less than the capabilities of a drilling rig. The workover begins by removing the wellhead and possibly the flow line then lifting the tubing hanger from the casing head and thus beginning to pull the completion out of the well. The string will almost always be fixed in place by at least one production packer. If the packer is retrievable it can be released easily enough and pulled out with the completion string. If it is permanent, then it is common to cut the tubing just above it and pull out the upper portion of the string. If necessary, the packer and the tubing left in hole can be milled out, though more commonly, the new completion will make use of it by setting a new packer just above it and running new tubing down to the top of the old.

Workovers on Casing

Although less exposed to wellbore fluids, casing strings too have been known to lose integrity. On occasion, it may be deemed economical to pull and replace it. Because casing strings are cemented in place, this is significantly more difficult and expensive than replacing the completion string. If in some instances the casing cannot be removed from the well, it may be necessary to sidetrack the offending area and recomplete, also an expensive process. For all but the most productive well, replacing casing would never be economical.

Subsea (Technology)

Subsea is a general term frequently used to refer to equipment, technology, and methods employed in marine biology, undersea geology, offshore oil and gas developments, underwater mining, and offshore wind power industries.

Oil and Gas

Oil and gas fields reside beneath many inland waters and offshore areas around the world, and in the oil and gas industry the term *subsea* relates to the exploration, drilling and development of oil and gas fields in underwater locations. Under water oil field facilities are generically referred to using a *subsea* prefix, such as *subsea well*, *subsea field*, *subsea project*, and *subsea development*.

Subsea oil field developments are usually split into *Shallow water* and *Deepwater* categories to distinguish between the different facilities and approaches that are needed. The term *shallow water* or *shelf* is used for shallow water depths where bottom-founded facilities like jackup drilling rigs and fixed offshore structures can be used, and where saturation diving is feasible.

Deepwater is a term often used to refer to offshore projects located in water depths greater than around 600 feet, where floating drilling vessels and floating oil platforms are used, and unmanned underwater vehicles are required as manned diving is not practical. Subsea completions can be traced back to 1943 with the Lake Erie completion at a 35-ft water depth. The well had a land-type christmas tree that required diver intervention for installation, maintenance, and flow line connections. Shell completed its first subsea well in the Gulf of Mexico in 1961; The first known subsea ultra-high pressure waterjet system capable of operating below 5,000 ft was developed in 2010 by Jet Edge and Chukar Waterjet. It was used to blast away hydrates that were clogging a containment system at the Gulf oil spill site.

Systems

Subsea production systems can range in complexity from a single satellite well with a flowline linked to a fixed platform, FPSO or an onshore installation, to several wells on a template or clustered around a manifold, and transferring to a fixed or floating facility, or directly to an onshore installation.

Subsea production systems can be used to develop reservoirs, or parts of reservoirs, which require drilling of the wells from more than one location. Deep water conditions, or even ultradeep water conditions, can also inherently dictate development of a field by means of a subsea production system, since traditional surface facilities such as on a steel-piled jacket, might be either technically unfeasible or uneconomical due to the water depth.

The development of subsea oil and gas fields requires specialised equipment. The equipment must be reliable enough to safe guard the environment, and make the exploitation of the subsea hydrocarbons economically feasible. The deployment of such equipment requires specialised and expensive vessels, which need to be equipped with diving equipment for relatively shallow equipment work (i.e. a few hundred feet water depth maximum), and robotic equipment for deeper water depths. Any requirement to repair or intervene with installed subsea equipment is thus normally very expensive. This type of expense can result in economic failure of the subsea development.

Subsea technology in offshore oil and gas production is a highly specialised field of application with particular demands on engineering and simulation. Most of the new oil fields are located in deepwater and are generally referred to as deepwater systems. Development of these fields sets strict requirements for verification of the various systems' functions and their compliance with current requirements and specifications. This is because of the high costs and time involved in changing a pre-existing system due to the specialised vessels with advanced onboard equipment. A full scale test (System Integration Test – SIT) does not provide satisfactory verification of deepwater systems because the test, for practical reasons, cannot be performed under conditions identical to those under which the system will later operate. The oil industry has therefore adopted modern data technology as a tool for virtual testing of deepwater systems that enables detection of costly faults at an early phase of the project. By using modern simulation tools models of deepwater systems can be set up and used to verify the system's functions, and dynamic properties, against various requirements specifications. This includes the model-based development of innovative high-tech plants and system solutions for the exploitation and production of energy resources in an environmentally friendly way as well as the analysis and evaluation of the dynamic behaviour of components and systems used for the production and distribution of oil and gas. Another part is the real-time virtual test of systems for subsea production, subsea drilling, supply above sea level, seismography, subsea construction equipment and subsea process measurement and control equipment.

Offshore Wind Power

The power transmission infrastructure for offshore wind power utilises a variety of subsea technologies for the installation and maintenance of submarine power transmission cables and other electrical energy equipment. In addition, the monopile foundations of fixed-bottom wind turbines and the anchoring and cable structures of floating wind turbines are regularly inspected with a variety of shipborne subsea technology.

Underwater Mining

Recent technological advancements have given rise to the use of remotely operated vehicles (ROVs) to collect mineral samples from prospective mine sites. Using drills and other cutting tools, the ROVs obtain samples to be analysed for desired minerals. Once a site has been located, a mining ship or station is set up to mine the area.

Remotely Operated Vehicles

Remotely Operated Vehicles (ROVs) are robotic pieces of equipment operated from afar to perform tasks on the sea floor. ROVs are available in a wide variety of function capabilities and complexities from simple "eyeball" camera devices, to multi-appendage machines that require multiple operators to operate or "fly" the equipment.

Organisations

A number of professional societies and trade bodies are involved with the subsea industry around the world. Such groups include

- Society for Underwater Technology
- Subsea Project Awards
- Subsea UK
- Subsea Valley The largest subsea technology business cluster.
- Society of Petroleum Engineers
- American Petroleum Institute (API)
- American Society of Mechanical Engineers (ASME)
- National Association of Corrosion Engineers (NACE).

Government agencies administer regulations in their territorial waters around the world. Examples of such government agencies are the Minerals Management Service (MMS, US), Norwegian Petroleum Directorate (NPD, Norway), and Health & Safety Executive (HSE, UK). The MMS administers the mineral resources in the US (using Code of Federal Regulations (CFR)) and provides management of all the US subsea mineral and renewable energy resources.

Well Kill

A well kill is the operation of placing a column of heavy fluid into a well bore in order to prevent the flow of reservoir fluids without the need for pressure control equipment at the surface. It works on the principle that the weight of the "kill fluid" or "kill mud" will be enough to suppress the pressure of the formation fluids. Well kills may be planned in the case of advanced interventions such as workovers, or be contingency operations. The situation calling for a well kill will dictate the method taken.Not all well kills are deliberate. Sometimes, the unintended build up of fluids, either from injection of chemicals like methanol from surface, or from liquids produced from the reservoir, can be enough to kill the well, particularly gas wells, which are notoriously easy to kill.

Principles

The principle of a well kill revolves around the weight of a column of fluid and hence the pressure exerted at the bottom.

$$P = hg\rho$$

Where P is the pressure at depth h in the column, g is the acceleration of gravity and ρ is the density of the fluid. It is common in the oil industry to use weight density, which is the product of mass density and the acceleration of gravity. This reduces the equation to:

$$P = h\gamma$$

Where γ is the weight density. Weight density may also be described as the pressure gradient because it directly determines how much extra pressure will be added by increasing depth of the column of fluid. The objective in a well kill, is to make the pressure at the bottom of the kill fluid equal (or slightly greater) than the pressure of the reservoir fluids.

Example

The pressure of the reservoir fluids at the bottom of the hole is 38MPa. We have a kill fluid with a weight density of 16kN.m^{-3}. What will need to be the height of the hydrostatic head in order to kill the well?

From the equation: $h = \dfrac{P}{\gamma}$

$$h = \frac{38MPa}{16kNm^{-3}}$$

$$h = 2375m$$

Therefore, a column of 2375m of this fluid is needed. It must be remembered that this refers to the true vertical depth of the column, not the measured depth, which is always larger than true vertical depth due to deviations from vertical.

Maths in the Oil Field

In the oil industry, a pure SI system is far from being used. Weight densities are commonly either given as specific gravity or in pounds per gallon. Simple conversion factors (0.433 for specific gravity and 0.052 for ppg) convert these values to a pressure gradient in psi per foot. Multiplying by the depth in feet gives the pressure at the bottom of the column.

Methods of Well Kill

Reverse Circulation

This is often tidiest way of making a pre-planned well kill. It involves pumping kill fluid down the 'A' annulus of the well, through a point of communication between it and the production tubing just above the production packer and up the tubing, displacing the lighter well bore fluids, which are allowed to flow to production.

The point of communication was traditionally a device called a sliding sleeve, or sliding side door, which is a hydraulically operated device, built into the production tubing. During normal operation, it would remain closed sealing off the tubing and the annulus, but for events such as this, it would be opened to allow the free flow of fluids between the two regions. These components have fallen out of favour as they were prone to leaking. Instead, it is now more common to punch a hole in the tubing for circulation kills. Although this permanently damages the tubing, given that most pre-planned well kills are for workovers, this is not an issue, since the tubing is being pulled for replacement anyway.

Bullheading

This is the most common method of a contingency well kill. If there is a sudden need to kill a well quickly, without the time for rigging up for circulation, the more blunt instrument of bullheading may be used.

This involves simply pumping the kill fluid directly down the well bore, forcing the well bore fluids back into the reservoir. This can be effective at achieving the central aim of a well kill; building up a sufficient hydrostatic head in the well bore. However, it can risk damaging the reservoir, by forcing undesired materials into it. The principal advantage is that it can be done with little advanced planning.

Forward Circulation

This is similar to reverse circulation, except the kill mud is pumped into the production tubing and circulated out through the annulus. Though effective, it is not as desirable since it is preferred that the well bore fluids be displaced out to production, rather than the annulus.

Lubricate and Bleed

This is the most time consuming form of well kill. It involves repeatedly pumping in small quantities of kill mud into the well bore and then bleeding off excess pressure. It works on the principle that the heavier kill mud will sink below the lighter well bore fluids and so bleeding off the pressure will remove the latter leaving an increasing quantity of kill mud in the well bore with successive steps.

Well Kills During Drilling Operations

During drilling, pressure control is maintained through the use of precisely concocted drilling fluid, which balances out the pressure at the bottom of the hole. In the event of suddenly encountering a high pressure pocket of, say gas (called a "kick"), it can become necessary to kill the well. This is done by pumping kill mud down the drill pipe, where it circulates out the bottom and into the well bore.

Reversing a Well Kill

The intention of a well kill (or the reality of an unintentional well kill) is to stop reservoir fluids flowing to surface. This of course creates problems when it is desirable to get the well flowing again. In order to reverse the well kill, the kill fluid must be displaced from the well bore. This involves injecting a gas at high pressure, usually nitrogen since it is inert and cheap. A gas can be put under sufficient pressure to allow it to push heavy kill fluid, but will then expand and become light once pressure is removed. This means that having displaced the kill fluid, it will not itself kill the well. The reservoir fluids should be able to flow to surface, displacing the gas.

The cheapest way to do it is similar to bullheading, where the nitrogen is pumped in under high pressure to force the kill fluid into the reservoir. This, of course, runs a high risk of causing well damage. The most effective way is to use coiled tubing, pumping the gas down the coil and circulating out the bottom into the well bore, where it will displace the kill mud to production.

Chapter 4

Semi-Submersible

A semi-submersible is a specialised marine vessel with good stability and seakeeping characteristics. The semi-submersible vessel design is commonly used in a number of specific offshore roles such as for offshore drilling rigs, safety vessels, oil production platforms and heavy lift cranes.

The terms *semisubmersible*, *semi-sub* or just *semi* are also generally used for this vessel design.

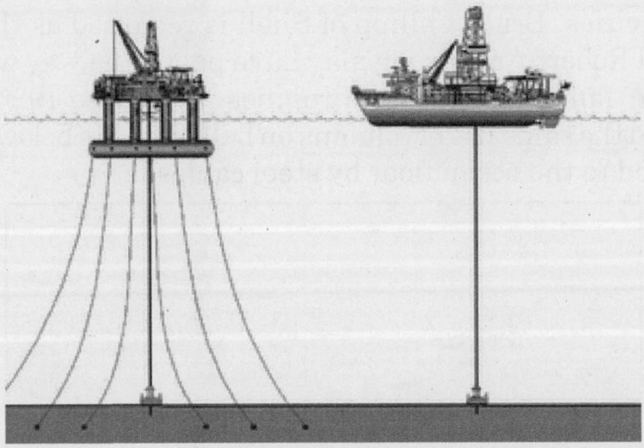

Figure: *Comparison of deepwater semi-submersible and drillship.*

Characteristics

Offshore drilling in water depth greater than around 120 metres requires that operations be carried out from a floating vessel, as fixed structures are not practical. Initially in the early 1950s monohull ships were used like CUSS I, but these were found to have significant heave, pitch and yaw motions in large waves, and the industry needed more stable drilling platforms.

A semi-submersible obtains its buoyancy from ballasted, watertight pontoons located below the ocean surface and wave action. The operating deck can be located high above the sea level due to the good stability of the concept, and therefore the operating deck is kept well away from the waves. Structural columns connect the pontoons and operating deck. With its hull structure submerged at a deep draft, the semi-submersible is less affected by wave loadings than a normal ship. With a small water-plane area, however, the semi-submersible is sensitive to load changes, and therefore must be carefully trimmed to maintain stability. Unlike a submarine or submersible, during normal operations, a semi-submersible vessel is never entirely underwater.

A semi-submersible vessel is able to transform from a deep to a shallow draft by deballasting (removing ballast water from the hull), and thereby become a surface vessel. The heavy lift vessels use this capability to submerge the majority of their structure, locate beneath another floating vessel, and then deballast to pick up the other vessel as a cargo.

Early History

The semi-submersible design was first developed for offshore drilling activities. Bruce Collipp of Shell is regarded as the inventor. But Edward Robert Armstrong may have paved the way with his idea of 'seadrome' landing strips for airplanes in the late 1920s, since his idea involved the same use of columns on ballast tanks below the surface and anchored to the ocean floor by steel cables.

Figure: *Blue Water Rig No. 1*

When oil drilling moved into offshore waters, fixed platform rigs and submersible rigs were built, but were limited to shallow waters. When demands for drilling equipment was needed in water depths greater than 100 feet (30 m) in the Gulf of Mexico, the first jackup rigs were built.

The first semisubmersible arrived by accident in 1961. Blue Water Drilling Company owned and operated the four column submersible drilling rig *Blue Water Rig No.1* in the Gulf of Mexico for Shell Oil Company. As the pontoons were not sufficiently buoyant to support the weight of the rig and its consumables, it was towed between locations at a draught mid way between the top of the pontoons and the underside of the deck. It was observed that the motions at this draught were very small and Blue Water Drilling and Shell jointly decided that the rig could be operated in the floating mode.

The first purpose built drilling semi-submersible *Ocean Driller* was launched in 1963. Since then, many semi-submersibles have been purpose-designed for the drilling industry mobile offshore fleet.

The industry quickly accepted the semi-submersible concept and the fleet increased rapidly to 30 units by 1972.

Applications

Mobile Offshore Drilling Units (MODU)

Semi-submersible rigs make stable platforms for drilling for offshore oil and gas. They can be towed into position by a tugboat and anchored, or moved by and kept in position by their own azimuth thrusters with dynamic positioning.

Drilling rig construction has historically occurred in boom periods and therefore 'batches' of drilling rigs have been built. Offshore drilling rigs have been classified in nominal 'generations' depending upon the year built and water depth capability as follows;

Generation	***Water Depth***		***Dates***
First	about 600 ft	200 m	Early 1960s
Second	about 1000 ft	300 m	1969–1974
Third	about 1500 ft	500 m	Early 1980s
Fourth	about 3000 ft	1000 m	1990's
Fifth	about 7500 ft	2500 m	1998–2004
Sixth	about 10000 ft	3000 m	2005–2010

The IMO MODU Code is an accredited design and operational guideline for Mobile Offshore Drilling Units of the semi-submersible type.

Semi-Submersible Crane Vessels (SSCV)

The advantages of the semi-submersible vessel stability were soon recognised for offshore construction when in 1978 Heerema Marine Contractors constructed the two sister crane vessels called *Balder* and *Hermod.* These semi-submersible crane vessels (SSCV) consist of two lower hulls (pontoons), three columns on each pontoon and an upper hull. Shortly after J. Ray McDermott and Saipem also introduced SSCV's, resulting in two new enormous vessels *DB-102* (now *Thialf*) and *Saipem 7000*, capable of lifting respectively 14,200 and 14,000 tons.

During transit an SSCV will be de-ballasted to a draught where only part of the lower hull is submerged. During lifting operations, the vessel will be ballasted down. This way, the lower hull is well submerged. This reduces the effect of waves and swell. High stability is obtained by placing the columns far apart. The high stability allows them to lift extreme high loads.

Offshore Support Vessels (OSV)

Semi-submersibles are particularly suited to a number of offshore support vessel roles because of their good stability, large deck areas, and variable deck load (VDL). Some of the most prominent vessels are;

- *Uncle John* – Diving / Construction support vessel, built for Houlder Offshore in 1977
- *Seaway Swan* – Diving Support Vessel built in 1977
- *Tharos* – Offshore Safety support vessel, built in 1979 and since converted into a drilling vessel, and rechristened *Transocean Marianas.*
- *Stadive* – Diving Support Vessel (DSV) built for Shell in 1982
- *Iolair* – Offshore safety support vessel, built for BP in 1982
- *Safe Karinia* – Offshore operations vessel, built in 1982
- *Polyconfidence* – Offshore accommodation platform, built in 1988.
- *Q4000* – Offshore Multiservice Vessel, built for Caldive in 2002
- *Ocean Odyssey* – Converted semi-submersible drilling rig used as a rocket launch pad.

Offshore Production Platforms

When oil fields were first developed in offshore locations, drilling semi-submersibles were converted for use as combined drilling and production platforms. These vessels offered very stable and cost effective platforms. The first semi-submersible floating production platform was the *Argyll FPF* converted from the *Transworld 58* drilling semi-submersible in 1975 for the Hamilton Brothers North Sea *Argyll* oil field.

As the oil industry has progressed into deeper water and harsh environments, purpose-built production semi-submersible platforms were designed. The first purpose-built semi-submersible production platform was for the Balmoral field, UK North Sea in 1986.

A summary of offshore semi-submersible oil production platforms is given in the following table derived from industry data.

Particularly large examples

The Petronius Platform is a compliant tower in the Gulf of Mexico, which stands 2,000 feet (610 m) above the ocean floor. It is one of the world's tallest structures.

The Hibernia platform in Canada is the world's largest (in terms of weight) offshore platform, located on the Jeanne D'Arc basin, in the Atlantic Ocean off the coast of Newfoundland. This *gravity base structure* (GBS), which sits on the ocean floor, is 364 feet (111 m) high and has storage capacity for 1.3 million barrels (210,000 m^3) of crude oil in its 278.8-foot (85.0 m) high caisson. The platform acts as a small concrete island with serrated outer edges designed to withstand the impact of an iceberg. The GBS contains production storage tanks and the remainder of the void space is filled with ballast with the entire structure weighing in at 1.2 million tons.

Royal Dutch Shell is currently developing the first Floating Liquefied Natural Gas (FLNG) facility, which will be situated approximately 200km off the coast of Western Australia and is due for completion around 2017. When finished, it will be the largest floating offshore facility. It is expected to be approximately 488m long and 74m wide with displacement of around 600,000t when fully ballasted.

Maintenance and Supply

A typical oil production platform is self-sufficient in energy and water needs, housing electrical generation, water desalinators and all of the equipment necessary to process oil and gas such that it can be either delivered directly onshore by pipeline or to a floating platform or tanker loading facility, or both. Elements in the oil/gas production process include wellhead, production manifold, production separator, glycol process to dry gas, gas compressors, water injection pumps, oil/gas export metring and main oil line pumps.

Larger platforms assisted by smaller ESVs (emergency support vessels) like the British Iolair that are summoned when something has gone wrong, *e.g.* when a search and rescue operation is required. During normal operations, PSVs (platform supply vessels) keep the platforms provisioned and supplied, and AHTS vessels can also supply them, as well as tow them to location and serve as standby rescue and firefighting vessels.

Crew

Essential Personnel

Not all of the following personnel are present on every platform. On smaller platforms, one worker can perform a number of different jobs. The following also are not names officially recognised in the industry:

- OIM (offshore installation manager) who is the ultimate authority during his/her shift and makes the essential decisions regarding the operation of the platform;
- operations team leader (OTL);
- offshore operations engineer (OOE) who is the senior technical authority on the platform;
- PSTL or operations coordinator for managing crew changes;
- dynamic positioning operator, navigation, ship or vessel maneuvering (MODU), station keeping, fire and gas systems operations in the event of incident;
- automation systems specialist, to configure, maintain and troubleshoot the process control systems (DCS), process safety systems, emergency support systems and vessel management systems;
- second mate to meet manning requirements of flag state, operates fast rescue craft, cargo operations, fire team leader;
- third mate to meet manning requirements of flag state, operate fast rescue craft, cargo operations, fire team leader;
- ballast control operator to operate fire and gas systems;
- crane operators to operate the cranes for lifting cargo around the platform and between boats;
- scaffolders to rig up scaffolding for when it is required for workers to work at height;
- coxswains to maintain the lifeboats and manning them if necessary;
- control room operators, especially FPSO or production platforms;

- catering crew, including people tasked with performing essential functions such as cooking, laundry and cleaning the accommodation;
- production techs to run the production plant;
- helicopter pilot(s) living on some platforms that have a helicopter based offshore and transporting workers to other platforms or to shore on crew changes;
- maintenance technicians (instrument, electrical or mechanical).

Incidental Personnel

Drill crew will be on board if the installation is performing drilling operations. A drill crew will normally comprise:

- Toolpusher
- Driller
- Roughnecks
- Roustabouts
- Company man
- Mud engineer
- Derrickhand
- Geologist.

Well services crew will be on board for well work. The crew will normally comprise:

- Well services supervisor
- Wireline or coiled tubing operators
- Pump operator.

Drawbacks

Risks

The nature of their operation — extraction of volatile substances sometimes under extreme pressure in a hostile environment — means risk; accidents and tragedies occur regularly. The U.S. Minerals Management Service reported 69 offshore deaths, 1,349 injuries, and 858 fires and explosions on offshore rigs in the Gulf of Mexico from 2001 to 2010. In July 1988, 167 people died when Occidental Petroleum's Piper Alpha offshore production platform, on the Piper field in the UK sector of the North Sea, exploded after a gas leak. The resulting investigation conducted by Lord Cullen and publicized in the first Cullen Report was highly critical of a number of areas, including, but not limited to, management within the company, the design of the structure, and the Permit to Work System. The report was commissioned in 1988, and was delivered November 1990. The accident greatly accelerated the practice of providing living accommodations on separate platforms, away from those used for extraction.

However, this was in itself a hazardous environment. In March 1980, the 'flotel' (floating hotel) platform *Alexander L. Kielland* capsized in a storm in the North Sea with the loss of 123 lives.

In 2001, *Petrobras 36* in Brazil exploded and sank five days later, killing 11 people. Given the number of grievances and conspiracy theories that involve the oil business, and the importance of gas/oil platforms to the economy, platforms in the United States are believed to be potential terrorist targets. Agencies and military units responsible for maritime counterterrorism in the US (Coast Guard, Navy SEALs, Marine Recon) often train for platform raids. On April 21, 2010, the *Deepwater Horizon* platform, 52 miles off-shore of Venice, Louisiana, (property of Transocean and leased to BP) exploded, killing 11 people, and sank two days later. The resulting undersea gusher, conservatively estimated to exceed 20 million US gallons (76,000 m^3) as of early June, 2010, became the worst oil spill in US history, eclipsing the Exxon Valdez oil spill.

Ecological Effects

In British waters, the cost of removing all platform rig structures entirely was estimated in 1995 at £1.5 billion, and the cost of removing all structures including pipelines—called a "clean sea" approach—at £3 billion. In the United States, Marine Biologist Milton Love has proposed that oil platforms off the California coast be retained as artificial reefs, instead of being dismantled (at great cost), because he has found them to be havens for many of the species of fish which are otherwise declining in the region, in the course of 11 years of research. Love is funded mainly by government agencies, but also in small part by the California Artificial Reef Enhancement Program. NOAA has said it is considering this course of action, but wants money to study the effects of the rigs in detail. Divers have been used to assess the fish populations surrounding the platforms. In the Gulf of Mexico, more than 200 platforms have been similarly converted.

Deepest Oil Platforms

The world's deepest oil platform is the floating Perdido which is a spar platform in the Gulf of Mexico in a water depth of 2,438 metres (7,999 ft). Non-floating compliant towers and fixed platforms, by water depth:

- Petronius Platform, 531 m (1,742 ft)
- Baldpate Platform, 502 m (1,647 ft)
- Bullwinkle Platform, 413 m (1,355 ft)
- Pompano Platform, 393 m (1,289 ft)
- Benguela-Belize Lobito-Tomboco Platform, 390 m (1,280 ft)
- Tombua Landana Platform, 366 m (1,201 ft)
- Harmony Platform, 366 m (1,201 ft)
- Troll A Platform, 303 m (994 ft)
- Gulfaks C Platform, 217 m (712 ft)

Pumpjack

A pumpjack (nodding donkey, pumping unit, horsehead pump, beam pump, sucker rod pump (SRP), grasshopper pump, thirsty bird, jack pump) is the overground drive for a reciprocating piston pump in an oil well.

It is used to mechanically lift liquid out of the well if there is not enough bottom hole pressure for the liquid to flow all the way to the surface. The arrangement is commonly used for onshore wells producing little oil. Pumpjacks are common in oil-rich areas.

Depending on the size of the pump, it generally produces 5 to 40 litres of liquid at each stroke. Often this is an emulsion of crude oil and water. Pump size is also determined by the depth and weight of the oil to remove, with deeper extraction requiring more power to move the heavier lengths of sucker rods.

A pumpjack converts the rotary mechanism of the motor to a vertical reciprocating motion to drive the pump shaft, and is exhibited in the characteristic nodding motion. The engineering term for this type of mechanism is a walking beam. It was often employed in stationary and marine steam engine designs in the 18th and 19th centuries.

Above Ground

In the early days, pumpjacks were actuated by rod lines running horizontally above the ground to an eccentric wheel in a mechanism known as a Central Power. The Central Power itself, which might operate a dozen or more pumpjacks, was powered by a steam or internal combustion engine or by an electric motor. Among the difficulties with this scheme was keeping the system balanced as individual well loads changed. Modern pumpjacks are powered by a prime mover. This is commonly an electric motor, but combustion engines are used in isolated locations without economic access to electricity. Common off-grid pumpjack engines run on casing gas produced from the well, but pumpjacks have been run on many types of fuel, such as propane and diesel. In harsh climates such motors and engines may be housed inside a shack to protect them from the elements.

Figure: *A pumpjack in California, USA*

The prime mover of the pumpjack runs a set of pulleys to the transmission which in turn drives a pair of cranks, generally with counterweights on them to assist the motor in lifting the heavy string of rods. The cranks in turn raise and lower one end of an I-beam which is free to move on an A-frame. On the other end of the beam, there is a curved metal box called a Horse Head or Donkeys Head, named so due to its appearance.

A cable made of steel (or, occasionally, fibreglass) called a bridle, connects the horse head to the polished rod, a piston that passes through the stuffing box. The polished rod has a close fit to the stuffing box, letting it move in and out of the tubing without fluid escaping. (The tubing is a pipe that runs to the bottom of the well through which the liquid is produced.) The bridle follows the curve of the horse head as it lowers and raises to create an almost completely vertical stroke. The polished rod is connected to a long string of rods called sucker rods, which run through the tubing all the way to the down-hole pump, usually positioned near the bottom of the well.

Down-Hole

At the bottom of the tubing is the down-hole pump. This pump has two ball check valves: a stationary valve at bottom called the standing valve, and a valve on the piston connected to the bottom of the sucker rods that travels up and down as the rods reciprocate, known as the travelling valve. Reservoir fluid enters from the formation into the bottom of the borehole through perforations that have been made through the casing and cement (the casing is a larger metal pipe that runs the length of the well, which has cement placed between it and the earth; the tubing, pump and sucker rods are all inside the casing). When the rods at the pump end are travelling up, the travelling valve is closed and the standing valve is open (due to the drop in pressure in the pump barrel). Consequently, the pump barrel fills with the fluid from the formation as the travelling piston lifts the previous contents of the barrel upwards. When the rods begin pushing down, the travelling valve opens and the standing valve closes (due to an increase in pressure in the pump barrel). The travelling valve drops through the fluid in the barrel (which had been sucked in during the upstroke). The piston then reaches the end of its stroke and begins its path upwards again, repeating the process. Often, gas is produced through the same perforations as the oil. This can be problematic if gas enters the pump, because it can result in what is known as gas locking, where insufficient pressure builds up in the pump barrel to open the valves (due to compression of the gas) and little or nothing is pumped. To preclude this, the inlet for the pump can be placed below the perforations. As the gas-laden fluid enters the well bore through the perforations, the gas bubbles up the annulus (the space between the casing and the tubing) while the liquid moves down to the standing valve inlet. Once at the surface, the gas is collected through piping connected to the annulus.

Water Well Pumpjacks

Pumpjacks can also be used to drive what would now be considered old-fashioned hand-pumped water wells. The scale of the technology is smaller than for an oil well, and can typically fit on top of an existing hand-pumped well head. The technology is simple, typically using a parallel-bar double-cam lift driven from a very low horsepower electric motor.

Although the flow rate for a water well pumpjack is low compared to a jet pump and the lifted water is not pressurized, the water well pumpjack has the option of hand pumping in an emergency, by hand-rotating the pumpjack cam to its lowest position, and attaching a manual handle to the top of the wellhead rod.

Well Stimulation

Well stimulation is a well intervention performed on an oil or gas well to increase production by improving the flow of hydrocarbons from the drainage area into the well bore.

Cleaning the Formation

The assortment of chemicals pumped down the well during drilling and completion can often cause damage to the surrounding formation by entering the reservoir rock and blocking the pore throats (the channels in the rock throughout which the reservoir fluids flow). Similarly, the act of perforating can have a similar effect by jetting debris into the perforation channels. Both these situations reduce the permeability in the near well bore area and so reduce the flow of fluids into the well bore.A simple and cheap solution is to pump chemicals from surface into the well to dissolve the offending material. Once dissolved, permeability should be restored and the reservoir fluids will flow into the well bore, cleaning up what is left of the damaging material. After initial completion, it is common to use concentrated formic acid to clean up any mud and skin damage. In this situation, the process is called acidizing.

In more serious cases, pumping from surface is insufficient as it does not target any particular location downhole and reduces the chances of the chemical retaining its effectiveness when it gets there. In these cases, it is necessary to spot the chemical directly at its target through the use of coiled tubing. Coiled tubing is run in hole with a jetting tool on the end. When the tool is at its target, the chemical is pumping through the pipe and is jetted directly onto the damaged area. This can be more effective than pumping from surface, though it is much more expensive, and accuracy is dependent on knowing the location of the damage. So well stimulation purpose is to remove damage and increase producting performance

Extending the Perforation Tunnels and Fractures

In cased hole completions, perforations are intended to create a hole through the steel casing so that the reservoir can be produced. The holes are typically formed by shaped explosives that perforate the casing and create a fractured hole into the reservoir rock for a short distance. In many cases, the tunnels created by the perforation guns do not provide enough surface area and it becomes desirable to create more area in contact with the wellbore.

In some cases, more area is needed if the reservoir is of low permeability. In other cases, damage caused by drilling and completion operations can be severe enough that the perforation tunnel does not effectively penetrate through the damaged volume near the bore. This means that the ability of fluids to flow into the existing perforation tunnels is too limited. One method to achieve more stimulation is by carrying out a hydraulic fracture treatment through the perforations.

If permeability is naturally low, then as fluid is drained from the immediate area, replacement fluid may not flow into the void sufficiently quickly to make up for the voidage and so the pressure drops. The well cannot then flow at a rate sufficient to make production economic. In this case, extending a hydraulic fracture deeper into the reservoir will allow higher production rates to be achieved.

Hydraulic fracturing is performed by injecting high pressure fluids into the wellbore and into the perforation tunnels to cause the rock formation to fracture. This can either be done by injecting hydraulic fluid from surface, a process called hydraulic fracturing or using an explosive to generate a high speed gas flow, a process called propellant stimulation.

Propellant stimulations can be a very economical way to clean up nearbore damage. Propellants are a low-explosive material that generate large amounts of gas downhole very rapidly. The gas pressure builds in the wellbore, increasing tension in the rock until it becomes greater than the breakdown pressure of the formation. Fracture length and fracture pattern are highly dependent on the type of propellant stimulation tool that is used.

Lifting the Well

Some stimulation techniques do not necessarily mean altering the permeability outside the well bore. Sometimes they involve making it easier for fluids to flow up the well bore having already entered. Gas lift is sometimes considered a form of stimulation, particularly when it is only used for starting up the well and shut off during steady state operation. More commonly though, lifting as a stimulation refers to trying to lift out heavy liquids that have accumulated at the bottom, either through water entry from the formation or through chemicals injected from surface such as scale inhibitors and methanol (hydrate inhibitor). These liquids sit at the bottom of the well as can act as a weight holding back the flow of reservoir fluids, essentially acting to kill the well. They can be removed by circulating nitrogen using coiled tubing.

Squeeze Job

Squeeze job, or squeeze cementing is a term often used in the oilfield to describe the process of injecting cement slurry into a zone, generally for pressure-isolation purposes. The term probably originated from the concept that enough water is "squeezed" out of the slurry to render it unflowable, so the portion that has actually entered the zone will stay in place when the squeeze pressure is removed. After surface indications (e.g., pressure reaching a predetermined maximum) that a squeeze has been attained, any still-pumpable cement slurry remaining in the drill pipe or tubing ideally can be reverse circulated out before it sets.

Usually the zone to be squeezed is isolated from above with a packer (and possibly from below with a bridge plug), but sometimes the squeezing pressure is applied to the entire casing string in what is known as a bradenhead squeeze, (named for an old manufacturer of casing heads).

Even if a drilling rig is on location, pumping operations usually are done by a service company's cementing unit that can easily mix small batches of cement slurry, measure displacement volume accurately to spot the slurry on bottom, then pump at very low rates and high pressures during the squeeze itself, and finally measure volumes accurately again when reversing out any excess slurry. A squeeze manifold is a compact arrangement of valves and pressure gauges that allows monitoring of the drill pipe and casing pressures throughout the job, and facilitates quick switching of the pumping pressure to either side while the fluid returning from the other side of well is directed to the mud pit or a disposal pit or tank.

The generic term "squeeze" also can apply to injection of generally small volumes of other liquids (e.g., treating fluids) into a zone under pressure. Bullhead squeeze (or just plain bullheading) refers to pumping kill-weight mud down the casing beneath closed blowout preventers in a kick-control situation when it isn't feasible to circulate in such from bottom.

Integrated Asset Modelling

Integrated Asset Modelling (or IAM) is the generic term used in the oil industry for computer modelling of both the subsurface and the surface elements of a field development. Historically the reservoir has always been modelled separately from the surface network and the facilities. In order to capture the interaction between those two or more standalone models, several time consuming iterations were required. For example a change in the water breakthrough leads to a change in the deliverability of the surface network which in turn leads to a production acceleration or deceleration in the reservoir. In order to go through this lengthy process more quickly the industry has slowly been adopting a more integrated approach which captures the constraints imposed by the infrastructure on the network immediately.

Basis

As the aim of an IAM is to provide a production forecast which honours both the physical realities of the reservoir and the infrastructure it needs to contain the following elements;

Typical Components of an IAM:

- a pressure network
- a subsurface saturation model
- an availability model
- a constraint manager
- a production optimisation algorithm.

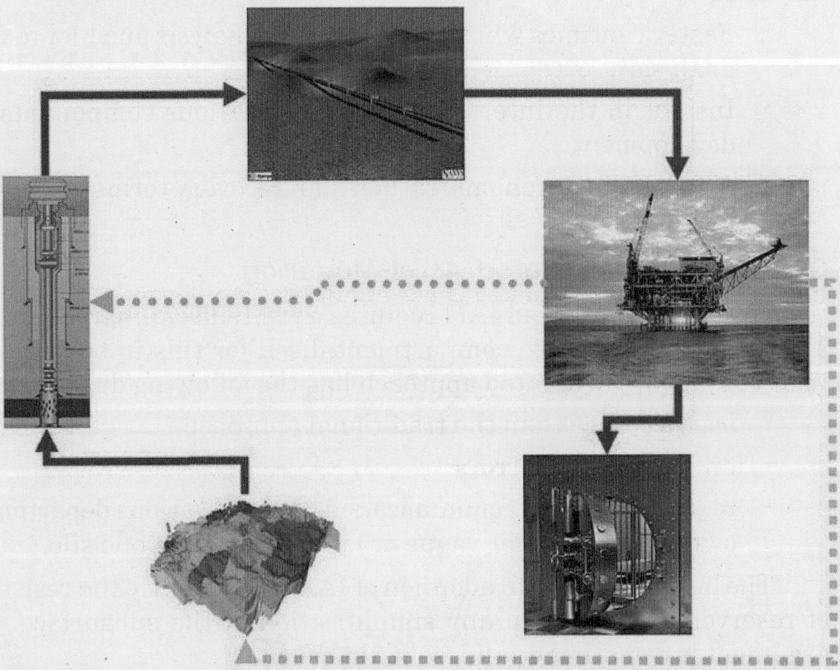

IPM vs. IAM

Historically the terms *Integrated Production Modelling* and *Integrated Asset Modelling* have been used interchangeably. The modern use of *Integrated Production Modelling* was coined when Petroleum Experts, Ltd. joined their MBAL modelling software with their GAP and Prosper modelling software to form an *Integrated Production Model.*

The phrase was coined unconsciously by an informal team of expert users in the US who worked on various projects together and separately using the products from Petroleum Experts, Ltd., a UK based company. The informal team members were Mike Arnondin (Mobil Oil), Mike Guardia (Michael A. Guardia & Associates), Allen Klingensmith (RDS), and Jeff Sieler (Shell). The term caught on quickly in the industry and was then formally adopted by Petroleum Experts, Ltd. British Petroleum (BP) decided to create their own term for internal use and hence arrived at *Integrated Asset Modelling* (IAM). The IAM term came about from a joint training course that BP developed along with Michael A. Guardia & Associates. IAM being coined by Simon Bishop (BP).

Benefits of Integrated Asset Modelling

Having an IAM built of an asset or future project offers several advantages;

- Faster runtimes which allow scenario analysis and Monte Carlo analysis
- Insight in the interactions between various components of a development
- An answer in economic rather than recovery terms (not always available).

Drawbacks of Integrated Asset Modelling

By its very nature an IAM requires a multi disciplinary approach. Most companies are too compartmentalised for this to be easy, as a result of this an integrated approach has the following drawbacks;

- less accurate than the traditional approach
- difficult to spot errors
- requires constant communication between various departments, ownership is either vague or too much part of one silo.

The biggest barrier to adoption of IAM is frequently the resistance of reservoir engineers to any simplification of the subsurface. This argument is sometimes valid, sometimes not.

Appropriate Use of IAM

As with any other software because of the inherent limitations in any virtual model use of an IAM is only appropriate during various stages of a project life. There are no hard and fast rules for this as there are a variety of software packages on the market which offer very accurate modelling of a very small scope to very rough modelling of a very large scope and anything in between. Currently the definition of IAM contains anything from daily optimisation to portfolio management. The success or failure of an IAM implementation project therefore depends on selecting the tool which is as complex as it needs to be but no more.

The following contains some examples of areas where an IAM is the appropriate decision support tool:

- Concept Select
- Debottlenecking and optimisation of very large or complex infrastructures
- Life of field analysis of production optimisation scenarios.

Note that for most of these areas the accuracy of the reservoir proxy is not important, the decision is made based on relative performance differences, not absolute values.

Approach

Several different software packages are commercially available and there is a clear difference in philosophy between some of them.

Linked Existing Software

Some vendors who have previously marketed standalone software for the subsurface and the surface are now marketing additional software which provides a datalink between the various packages. The obvious benefit of this approach is that there is no loss in accuracy and it does not require a remodelling exercise. However this approach also has its drawbacks, there is no time gain and the integration component of the entire package requires expertise which is not readily available, external specialist are frequently called upon to build and maintain the links between the components.

Bespoke Software

There are relatively few software packages on the market which are truly integrated. None of these can match the accuracy of the linked existing software but they offer the benefit of shorter runtimes and lower expertise tresholds.

Software as a Service

A number of the established service companies now offer integrated asset modelling as a service. In practice this means that existing models will be either converted or linked by specialists to form an integrated solution. This solution is expensive but frequently the preferred option if the highest accuracy is required.

Petrophysics

Petrophysics (petro is Latin for "rock" and physics is the study of nature) is the study of the physical and chemical properties that describe the occurrence and behaviour of rocks, soils and fluids. Petrophysics mainly studies reservoirs of resources, including ore deposits and oil or natural gas reservoirs. Petrophysicists in the oil and gas industry typically are employed in helping the engineers and other geoscientists understand the rock properties of the reservoir. Petrophysicists evaluate the reservoir rock properties by employing well log measurements, in which a string of measurement tools are inserted in the borehole, core measurements, in which rock samples are retrieved from subsurface, and sometimes seismic measurements, and combining them with geology and geophysics.

Petrophysical studies are used by petroleum engineering, geology, mineralogy, exploration geophysics and other related studies.

Some of the key properties studied in petrophysics are lithology, porosity, water saturation, permeability, density, solid mechanics, magnetization, electrical conductivity, thermal conductivity and radioactivity.

While most petrophysicists work in the oil and gas industry, some also work in the mining and water resource industries. The properties measured or computed fall into three broad categories: conventional petrophysical properties, rock mechanical properties, and ore quality.

Conventional Petrophysical Properties

Most petrophysicists are employed to compute what are commonly called conventional (or reservoir) petrophysical properties. These are:

Lithology: What type of rock is it. When combined with local geology and core study, geoscientists can use log measurements such as natural gamma, neutron, density, Photoelectric, resistivity or their combination to determine the lithology downhole.

Porosity: φ The amount of pore (or fluid occupied) space in the rock. This is typically measured using an instrument that measures the reaction of the rock to bombardment by neutrons or by gamma rays. Sonic wave speed and NMR logs are also measured to derive rock porosity.

Water saturation: S_w The fraction of the pore space occupied by water. This is typically measured using an instrument that measures the resistivity of the rock.

Permeability: The quantity of fluid (usually hydrocarbon) that can flow from the rock as a function of time and pressure. Formation testing is so far the only direct downhole permeability measurement. In case of its absence, which is common in most cases, permeability estimation is derived from other measurements, such as porosity, NMR, sonic, by empirical correlations.

Thickness of rock with enough permeability to deliver fluids to a well bore. This property is often called "Net reservoir rock." In the oil and gas industry, another quantity "Net Pay" is computed which is the thickness of rock that can deliver hydrocarbons to the well bore at a profitable rate.

Reservoir models are built upon their measured and derived properties to estimate the amount of hydrocarbon present in the reservoir, the rate at which that hydrocarbon can be produced to the Earth's surface through wellbores and the fluid flow in rocks. In the water resource industry, similar models are used to compute how much water can be produced to the surface over long periods of time, without depleting the aquifer.

Rock Mechanical Properties

Some petrophysicists use acoustic and density measurements of rocks to compute their mechanical properties and strength. They measure the compressional (P) wave velocity of sound through the rock and the shear (S) wave velocity and use these with the density of the rock to compute: The rocks compressive strength which is the compressive stress that causes a rock to fail. The rocks flexibility, the relationship between stress and deformation for a rock. Converted-wave analysis is also used to determine subsurface lithology and porosity. These measurements are useful to design programs to drill wells that produce oil and gas. The measurements are also used to design dams, roads, foundations for buildings and many other large construction projects. They can also be used to help interpret seismic signals from the Earth, either man-made seismic signals or those from earthquakes.

Ore Quality

Bore holes can be drilled into ore bodies (for example coal seams or gold ore) and either rock samples taken to determine the ore or coal quality at each bore hole location or the wells can be wireline logged to make measurements that can be used to infer quality. Some petrophysicists do this sort of analysis. The information is mapped and used to make mine development plans.

Methods of Analysis

Coring and core analysis is a direct measurement of petrophysical properties. In the petroleum industry rock samples are retrieved from subsurface and measured by core labs of oil company or some commercial core measurement service companies. This process is time consuming and expensive, thus can not be applied to all the wells drilled in a field.

Well Logging is used as a relatively inexpensive method to obtain petrophysical properties downhole. Measurement tools are conveyed downhole using either wireline or LWD method.

The first "track", shows the natural gamma radiation level of the rock. The gamma radiation level "log" shows increasing radiation to the right and decreasing radiation to the left. The rocks emitting less radiation have more yellow shading.

The detector is very sensitive and the amount of radiation is very low. In clastic rock formations, rocks that have smaller amounts of radiation are more likely to be coarser grained and have more pore space, rocks with higher amounts of radiation are more likely to have finer grains and less pore space (Poupon, Clavier, Dumanoir, Gaymard, and Misk, Journal of Petroleum Technology, July, 1970, page 868).

The second track over in the plot records the depth below the reference point which is usually the Kelly bush or rotary table in feet, so these rocks are 11,900 feet below the surface of earth.

In the third track, the electrical resistivity of the rock is presented. The water in this rock is salty and the salt in the water causes the water to be electrically conductive such that lower resistivity is caused by increasing water saturation and decreasing hydrocarbon saturation (Brown, A Mathematical Comparison of Common Saturation Equations, SPWLA twenty-seventh annual logging symposium, June, 1986, paper T). The fourth track, shows the computed water saturation, both as "total" water (including the water bound to the rock) in magenta and the "effective water" or water that is free to flow in black. Both quantities are given as a fraction of the total pore space.

The fifth track shows the fraction of the total rock that is pore space, filled with fluids. The display of the pore space is divided into green for oil and blue for movable water. The black line shows the fraction of the pore space which contains either water or oil that can move, or be "produced." In addition to the what is included in blackline, the magenta line includes the water that is permanently bound to the rock.

The last track is a representation of the solid portion of the rock. The yellow pattern represents the fraction of the rock (excluding fluids) that is composed of coarser grained sandstone. The gray pattern represents the fraction of rock that is composed of finer grained "shale." The sandstone is the part of the rock that contains the producible hydrocarbons and water.

Rock Volumetric Model for Shaly Sand Formation

The following definition and petrophysics model are a typical shaly sand formation model which assumes: 1. Shale is composed of silt, clay and their bounded water which will not flow. 2. Hydrocarbon are stored only in pore space in sand matrix.

ΦT - Total porosity (PHIT), which includes the pore space in sand and shale.

Sw - Total water saturation, the fraction of the pore space occupied by water.

Φe - Effective shale corrected porosity which includes only the pore space in sand. The pore space in shale which is filled with bounded water is excluded.

Swe - Effective shale corrected water saturation. The volumetric fraction of ?e which is occupied by water.

Vsh - Volumetric fraction of shale. This includes medium to very fine silt plus clay and the shale bound water.

Φsh - Shale porosity. Volumetric fraction of pore space in shale. These pore space is filled with bounded water by definition.

Key equations:

$$(1-\Phi e-Vsh) + Vsh + \Phi e*Swe + \Phi e*(1-Swe) = 1$$

Sandstone matrix volume + shale volume + water volume in sand + hydrocarbon volume in sand = total rock volume

$$\Phi e = \Phi T - Vsh *\Phi sh$$

Petroleum Geology

Petroleum geology refers to the specific set of geological disciplines that are applied to the search for hydrocarbons (oil exploration).

Sedimentary Basin Analysis

Petroleum geology is principally concerned with the evaluation of seven key elements in sedimentary basins:

A structural trap, where a fault has juxtaposed a porous and permeable reservoir against an impermeable seal. Oil (shown in red) accumulates against the seal, to the depth of the base of the seal. Any further oil migrating in from the source will escape to the surface and seep. In general, all these elements must be assessed via a limited 'window' into the subsurface world, provided by one (or possibly more) exploration wells. These wells present only a 1-dimensional segment through the Earth and the skill of inferring 3-dimensional characteristics from them is one of the most fundamental in petroleum geology. Recently, the availability of inexpensive, high quality 3D seismic data (from reflection seismology) and data from various electromagnetic geophysical techniques (such as Magnetotellurics) has greatly aided the accuracy of such interpretation. The following section discusses these elements in brief.

Evaluation of the source uses the methods of geochemistry to quantify the nature of organic-rich rocks which contain the precursors to hydrocarbons, such that the type and quality of expelled hydrocarbon can be assessed. The reservoir is a porous and permeable lithological unit or set of units that holds the hydrocarbon reserves. Analysis of reservoirs at the simplest level requires an assessment of their porosity (to calculate the volume of in situ hydrocarbons) and their permeability (to calculate how easily hydrocarbons will flow out of them). Some of the key disciplines used in reservoir analysis are the fields of structural analysis, stratigraphy, sedimentology, and reservoir engineering.

The seal, or cap rock, is a unit with low permeability that impedes the escape of hydrocarbons from the reservoir rock. Common seals include evaporites, chalks and shales. Analysis of seals involves assessment of their thickness and extent, such that their effectiveness can be quantified.

The trap is the stratigraphic or structural feature that ensures the juxtaposition of reservoir and seal such that hydrocarbons remain trapped in the subsurface, rather than escaping (due to their natural buoyancy) and being lost. Analysis of maturation involves assessing the thermal history of the source rock in order to make predictions of the amount and timing of hydrocarbon generation and expulsion.

Finally, careful studies of migration reveal information on how hydrocarbons move from source to reservoir and help quantify the source (or kitchen) of hydrocarbons in a particular area.

Major Subdisciplines in Petroleum Geology

Several major subdisciplines exist in petroleum geology specifically to study the seven key elements discussed above.

Analysis of Source Rocks

In terms of source rock analysis, several facts need to be established. Firstly, the question of whether there actually is any source rock in the area must be answered. Delineation and identification of potential source rocks depends on studies of the local stratigraphy, palaeogeography and sedimentology to determine the likelihood of organic-rich sediments having been deposited in the past.

If the likelihood of there being a source rock is thought to be high, the next matter to address is the state of thermal maturity of the source, and the timing of maturation. Maturation of source rocks depends strongly on temperature, such that the majority of oil generation occurs in the 60° to 120°C range. Gas generation starts at similar temperatures, but may continue up beyond this range, perhaps as high as 200°C. In order to determine the likelihood of oil/gas generation, therefore, the thermal history of the source rock must be calculated. This is performed with a combination of geochemical analysis of the source rock (to determine the type of kerogens present and their maturation characteristics) and basin modelling methods, such as back-stripping, to model the thermal gradient in the sedimentary column.

Basin Analysis

A full scale basin analysis is usually carried out prior to defining leads and prospects for future drilling. This study tackles the petroleum system and studies source rock (presence and quality); burial history; maturation (timing and volumes); migration and focus; and potential regional seals and major reservoir units (that define carrier beds). All these elements are used to investigate where potential hydrocarbons might migrate towards. Traps and potential leads and prospects are then defined in the area that is likely to have received hydrocarbons.

Exploration Stage

Although a basin analysis is usually part of the first study a company conducts prior to moving into an area for future exploration, it is also sometimes conducted during the exploration phase. Exploration geology comprises all the activities and studies necessary for finding new hydrocarbon occurrence. Usually seismic (or 3D seismic) is shot and old exploration data (seismic lines, well logs, reports) are used to expand upon. Sometimes gravity and magnetic studies are conducted and oil seeps and spills are mapped to find potential areas for hydrocarbon occurrences. As soon as a significant hydrocarbon occurrence is found by an exploration- or wildcat-well the appraisal stage is set in.

Appraisal Stage

The Appraisal stage is used to delineate the extent of the discovery. Furthermore reservoir properties, connectivity, hydrocarbon type and gas-oil and oil-water contacts are determined to calculate potential recoverable volumes. This is usually done by drilling more appraisal wells around the initial exploration well. Furthermore some production tests may give insight in reservoir pressures and connectivity. While geochemical and petrophysical analysis gives information on the type (viscosity, chemistry, API, carbon content, etc) of the hydrocarbon and the nature of the reservoir (porosity, permeability, etc).

Production Stage

After a hydrocarbon occurrence has been discovered and appraisal has indicated it is a commercial find the production stage is initiated. This stage focuses on extracting the hydrocarbons in a controlled way (without damaging the formation, within commercial favourable volumes, etc). Production wells are drilled and completed in strategic positions. 3D seismic is usually available by this stage to target wells precisely for optimal recovery. Sometimes enhanced recovery (steam injection, pumps, etc) is used to extract more hydrocarbons or to redevelop abandoned fields.

Analysis of Reservoir

The existence of a reservoir rock (typically, sandstones and fractured limestones) is determined through a combination of regional studies (i.e. analysis of other wells in the area), stratigraphy and sedimentology (to quantify the pattern and extent of sedimentation) and seismic interpretation. Once a possible hydrocarbon reservoir is identified, the key physical characteristics of a reservoir that are of interest to a hydrocarbon explorationist are its bulk rock volume, net-to-gross ratio, porosity and permeability.

Bulk rock volume, or the gross rock volume of rock above any hydrocarbon-water contact, is determined by mapping and correlating sedimentary packages. The net-to-gross ratio, typically estimated from analogues and wireline logs, is used to calculate the proportion of the sedimentary packages that contains reservoir rocks. The bulk rock volume multiplied by the net-to-gross ratio gives the net rock volume of the reservoir. The net rock volume multiplied by porosity gives the total hydrocarbon pore volume i.e. the volume within the sedimentary package that fluids (importantly, hydrocarbons and water) can occupy. The summation of these volumes for a given exploration prospect will allow explorers and commercial analysts to determine whether a prospect is financially viable. Traditionally, porosity and permeability were determined through the study of hand specimens, contiguous parts of the reservoir that outcrop at the surface and by the technique of formation evaluation using wireline tools passed down the well itself. Modern advances in seismic data acquisition and processing have meant that seismic attributes of subsurface rocks are readily available and can be used to infer physical/sedimentary properties of the rocks themselves.

Artificial Lift

Artificial lift refers to the use of artificial means to increase the flow of liquids, such as crude oil or water, from a production well. Generally this is achieved by the use of a mechanical device inside the well (known as pump or velocity string) or by decreasing the weight of the hydrostatic column by injecting gas into the liquid some distance down the well.

Artificial lift is needed in wells when there is insufficient pressure in the reservoir to lift the produced fluids to the surface, but often used in naturally flowing wells (which do not technically need it) to increase the flow rate above what would flow naturally. The produced fluid can be oil, water or a mix of oil and water, typically mixed with some amount of gas.

Usage

Any liquid-producing reservoir will have a 'reservoir pressure': some level of energy or potential that will force fluid (liquid, gas or both) to areas of lower energy or potential. The concept is similar to that of water pressure in a municipal water system. As soon as the pressure inside a production well is decreased below the reservoir pressure, the reservoir will act to fill the well back up, just like opening a valve on a water system. Depending on the depth of the reservoir and density of the fluid, the reservoir may or may not have enough potential to push the fluid to the surface - a deeper well or a heavier mixture results in a higher pressure requirement.

Most oil production reservoirs have sufficient potential to naturally produce oil and gas - which are light - during the early phases of production. Water - which is heavier than oil and much heavier than gas - will eventually encroach into production, and reservoir pressure will decrease as the reservoir depletes, naturally causing all wells to stop flowing. At some point, most well operators will implement an artificial lift plan to continue or increase production. Most water-producing wells, by contrast, will need artificial lift from the very beginning of production because they do not benefit from the lighter density of oil and gas.

Technologies

Hydraulic Pumping Systems: Hydraulic pumping systems transmit energy to the bottom of the well by means of pressurized power fluid that flows down in the wellbore tubular to a subsurface pump. There are two types of hydraulic subsurface pump:

1. a reciprocating piston pump, where one side is powered by the injected fluid while the other side pumps the produced fluids to surface
2. a jet pump, where the injected fluid passes through a nozzle creating a venturi effect pushing the produced fluids to surface.

These systems are very versatile and have been used in shallow depths (1000 ft) to deeper wells (18,000 ft), low rate wells with production in the tens of barrels per day to wells producing in excess of 10,000 barrels (1,600 m3) per day. Certain substances can be mixed in with the injected fluid to help deal or control with corrosion, paraffin and emulsion problems. Hydraulic pumping systems are also suitable for deviated wells where conventional pumps such as the rod pump are not feasible.

These systems have also some disadvantages. They are sensitive to solids and are the least efficient lift method. While typically the cost of deploying these systems has been very high, new coiled tubing umbilical technologies are in some cases greatly reducing the cost.

ESP

Electric Submersible Pumps consist of a downhole pump (a series of centrifugal pumps), an electrical motor which transforms the electrical power into kinetic energy to turn the pump, a separator or protector to prevent produced fluids from entering the electrical motor, and an electric power cable that connects the motor to the surface control panel. ESP is a very versatile artificial lift method and can be found in operating environments all over the world. They can handle a very wide range of flow rates (from 200 to 90,000 barrels (14,000 m3) per day) and lift requirements (from virtually zero to 10,000 ft (3,000 m) of lift). They can be modified to handle contaminants commonly found in oil, aggressive corrosive fluids such as H2S and CO2, and exceptionally high downhole temperatures. Increasing water cut has been shown to have no significant detrimental effect on the ESP performance. It is possible to locate them in vertical, deviated, or horizontal wells, but it is recommended to deploy them in a straight section of casing for optimum run life performance.

Although latest developments are aimed to enhance the ESP capabilities to handle gas and sand, they still need more technological development to avoid gas locks and internal erosion. Until recently, ESPs have come with an often prohibitive price tag due to the cost of deployment which can be in excess of $20,000.

Gas Lift

Gas lift is another widely used artificial lift method. As the name denotes, gas is injected in the tubing to reduce the weight of the hydrostatic column, thus reducing the back pressure and allowing the reservoir pressure to push the mixture of produce fluids and gas up to the surface. The gas lift can be deployed in a wide range of well conditions (from 30,000 bbl/d (4,800 m3/d) to 15,000 ft (4,600 m)). Gas lifts can cope well with abrasive elements and sand, and the cost of workover is minimum. Gas lifted wells are equipped with side pocket mandrels and gas lift injection valves. This arrangement allows a deeper gas injection in the tubing. The gas lift system has some disadvantages. There has to be a source of gas, some flow assurance problems such as hydrates can be triggered by the gas lift.

PCP

Progressing Cavity Pumps, PCP, are also widely applied in the oil industry. The PCP consists of a stator and a rotor. The rotor is rotated using either a top side motor or a bottom hole motor. The rotation created sequential cavities and the produced fluids are pushed to surface. The PCP is a flexible system with a wide range of applications in terms of rate(up to 5,000 bbl/d (790 m3/d) and 6,000 ft (1,800 m)). They offer outstanding resistance to abrasives and solids but they are restricted to setting depths and temperatures. Some components of the produced fluids like aromatics can also deteriorate the stator's elastomer.

Rod Pumps

Rod pumps are long slender cylinders with both fixed and moveable elements inside. The pump is designed to be inserted inside the tubing of a well and its main purpose is to gather fluids from beneath it and lift them to the surface. The most important components are: the barrel, valves (travelling and fixed) and the piston. It also has another 18 to 30 components which are called "fittings".

Components

Every part of the pump is important for its correct operation. The most commonly used parts are described below:

- Barrel: The barrel is a long cylinder, which can be from 10 to 36 feet (11 m) long, with a diametre of 1.25 inches (32 mm) to 3.75 inches (95 mm). After experience with several materials for its construction, the API (American Petroleum Institute) standardized the use of two materials or compositions for this part: carbon steel and brass, both with an inside coating of chrome. The advantage of brass against the harder carbon steel is its 100% resistance to corrosion.
- Piston/Plunger: This is a nickel-metal sprayed steel cylinder that goes inside the barrel. Its main purpose is to create a sucking effect that lifts the fluids beneath it and then, with the help of the valves, take the fluids above it, progressively, out of the well.
- Valves: The valves have two components - the seat and the ball - which create a complete seal when closed. The most commonly used seats are made of carbon nitride and the ball is often made of silicon nitride. In the past, balls of iron, ceramic and titanium were used. Titanium balls are still being used but only where crude oil is extremely dense and/or the quantity of fluid to be lifted is large. The most common configuration of a rod pump requires two valves, called the travelling valve and the fixed (or static or standing) valve.
- Piston rod: This is a rod that connects the piston with the outside of the pump. Its main purpose is to transfer the up/down reciprocating energy produced by the "Nodding Donkey" (pumping unit) installed above ground.
- Fittings: The rest of the parts of the pump are called fittings and are, basically, small pieces designed to keep everything hold together in the right place. Most of these parts are designed to let the fluids pass uninterrupted.
- Filter/Strainer: The job of the filter, as guessed, is to stop big parts of rock, rubber or any other garbage that might be loose in the well from being sucked into the pump. There are several types of filters, with the most common being an iron cylinder with enough holes in it to permit the entrance of the amount of fluid the pump needs.

Extraction of Petroleum

The extraction of petroleum is the process by which usable petroleum is extracted and removed from the earth.

Figure: *Pumpjack on an oil well in Texas*

Locating the Oil Field

Geologists use seismic surveys to search for geological structures that may form oil reservoirs. The "classic" method includes making an underground explosion nearby and observing the seismic response that provides information about the geological structures under the ground. However, "passive" methods that extract information from naturally-occurring seismic waves are also known.

Other instruments such as gravimetres and magnetometres are also sometimes used in the search for petroleum. Extracting crude oil normally starts with drilling wells into the underground reservoir. When an oil well has been tapped, a geologist (known on the rig as the "mudlogger") will note its presence. Such a "mudlogger" is known to be sitting on the rig. Historically, in the USA, some oil fields existed where the oil rose naturally to the surface, but most of these fields have long since been used up, except in certain places in Alaska. Often many wells (called multilateral wells) are drilled into the same reservoir, to ensure that the extraction rate will be economically viable. Also, some wells (secondary wells) may be used to pump water, steam, acids or various gas mixtures into the reservoir to raise or maintain the reservoir pressure, and so maintain an economic extraction rate.

Drilling

The oil well is created by drilling a hole into the earth with an oil rig. A steel pipe (casing) is placed in the hole, to provide structural integrity to the newly drilled wellbore. Holes are then made in the base of the well to enable oil to pass into the bore. Finally a collection of valves called a "Christmas Tree" is fitted to the top, the valves regulating pressures and controlling flows.

Oil Extraction and Recovery

Primary Recovery: During the primary recovery stage, reservoir drive comes from a number of natural mechanisms. These include: natural water displacing oil downward into the well, expansion of the natural gas at the top of the reservoir, expansion of gas initially dissolved in the crude oil, and gravity drainage resulting from the movement of oil within the reservoir from the upper to the lower parts where the wells are located. Recovery factor during the primary recovery stage is typically 5-15%.

While the underground pressure in the oil reservoir is sufficient to force the oil to the surface, all that is necessary is to place a complex arrangement of valves (the Christmas tree) on the well head to connect the well to a pipeline network for storage and processing.

Secondary Recovery

Over the lifetime of the well the pressure will fall, and at some point there will be insufficient underground pressure to force the oil to the surface. After natural reservoir drive diminishes, secondary recovery methods are applied. They rely on the supply of external energy into the reservoir in the form of injecting fluids to increase reservoir pressure, hence replacing or increasing the natural reservoir drive with an artificial drive. Sometimes pumps, such as beam pumps and electrical submersible pumps (ESPs), are used to bring the oil to the surface.

Other secondary recovery techniques increase the reservoir's pressure by water injection, natural gas reinjection and gas lift, which injects air, carbon dioxide or some other gas into the bottom of an active well, reducing the overall density of fluid in the wellbore. Typical recovery factor from water-flood operations is about 30%, depending on the properties of oil and the characteristics of the reservoir rock. On average, the recovery factor after primary and secondary oil recovery operations is between 35 and 45%.

Tertiary Recovery

Tertiary, or enhanced oil recovery methods increase the mobility of the oil in order to increase extraction.

Thermally enhanced oil recovery methods (TEOR) are tertiary recovery techniques that heat the oil, thus reducing its viscosity and making it easier to extract. Steam injection is the most common form of TEOR, and is often done with a cogeneration plant. In this type of cogeneration plant, a gas turbine is used to generate electricity and the waste heat is used to produce steam, which is then injected into the reservoir.

This form of recovery is used extensively to increase oil extraction in the San Joaquin Valley, which has very heavy oil, yet accounts for 10% of the United States' oil extraction. In-situ burning is another form of TEOR, but instead of steam, some of the oil is burned to heat the surrounding oil.

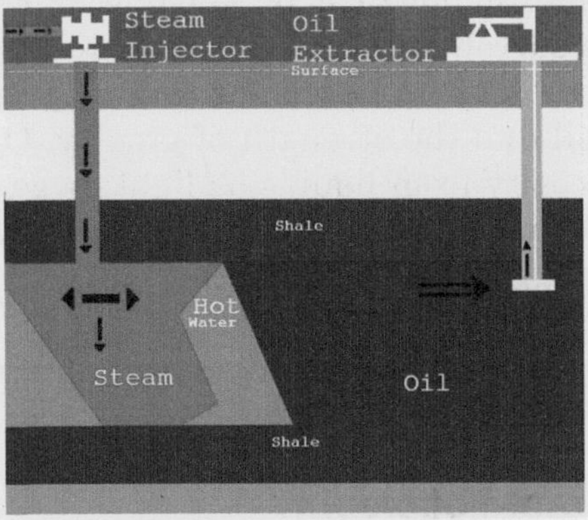

***Figure:** Steam is injected into many oil fields where the oil is thicker and heavier than normal crude oil*

Occasionally, surfactants (detergents) are injected to alter the surface tension between the water and oil in the reservoir, mobilising oil which would otherwise remain in the reservoir as residual oil.

Another method to reduce viscosity is carbon dioxide flooding.

Tertiary recovery allows another 5% to 15% of the reservoir's oil to be recovered.

Tertiary recovery begins when secondary oil recovery isn't enough to continue adequate extraction, but only when the oil can still be extracted profitably. This depends on the cost of the extraction method and the current price of crude oil. When prices are high, previously unprofitable wells are brought back into use and when they are low, extraction is curtailed.

Microbial treatments is another tertiary recovery method. Special blends of the microbes are used to treat and break down the hydrocarbon chain in oil thus making the oil easy to recover as well as being more economic versus other conventional methods. In some states, such as Texas, there are tax incentives for using these microbes in what is called a secondary tertiary recovery. Very few companies supply these, however companies like Bio Tech, Inc. have proven very successful in waterfloods across Texas.

Recovery Rates

The amount of oil that is recoverable is determined by a number of factors including the permeability of the rocks, the strength of natural drives (the gas present, pressure from adjacent water or gravity), and the viscosity of the oil. When the reservoir rocks are "tight" such as shale, oil generally cannot flow through but when they are permeable such as in sandstone, oil flows freely. The flow of oil is often helped by natural pressures surrounding the reservoir rocks including natural gas that may be dissolved in the oil, natural gas present above the oil, water below the oil and the strength of gravity. Oils tend to span a large range of viscosity from liquids as light as gasoline to heavy as tar. The lightest forms tend to result in higher extraction rates.

Petroleum engineering is the discipline responsible for evaluating which well locations and recovery mechanisms are appropriate for a reservoir and for estimating recovery rates and oil reserves prior to actual extraction.

Block (Extraction of Petroleum)

A block in the context of extraction of petroleum, (and natural gas) signifies a sub-division of an underground reservoir for the purpose of administering a certain production allowable.

Deep Well Drilling

Deep well drilling involves drilling (a well) to a desired "deep" depth of 10,000 ft or more.

Overview

Deep well drilling is sometimes more importantly characterised by the requirement of having the drill bit finally positioned within a specified radius of its surface starting location. This sometimes requires maneuvering the direction (azimuth) of the drill bit by various techniques (kicking) so as to arrive at the desired location at both the desired depth and within a specified radius of the starting point. The angle of drilling at any arbitrary azimuth can be derived by knowing the TVD (true vertical depth) and the MD (measured depth or hypotenuse) of any given segment.

Mathematics

This section demonstrates the formulae used to determine the current drill-bit location relative to its starting position given the azimuth, angle from vertical ("plumb angle") and either the TVD or MD of each drilling segment. If both the TVD and MD are known for any segment then the "plumb angle" for that segment is redundant.

This section also provides the formula for a "straight line" solution to arrive at an arbitrary depth and radial location given its current position.

Assume:

Target is 11,560 ft down Best drilling access is 432 feet away Azimuth of 38° on the surface.

So, you should set your azimuth direction at: (38° + 180°) = 218°

Then applying the rules of trigonometry:

The angle is defined by using the arctan function to the ratio of horizontal and vertical:

$$\arctan\frac{432}{11560} = 2.14°$$

If the drill bit doesn't get deflected then your target can be calculated by applying pythagorus' theorem:

$$\sqrt{(11560^2 + 432^2)} = 11568 ft.$$

Casing (Borehole)

Casing is large diametre pipe that is assembled and inserted into a recently drilled section of a borehole and typically held into place with cement.

Purpose

Casing that is cemented in place aids the drilling process in several ways:

- Prevent contamination of fresh water well zones.
- Prevent unstable upper formations from caving-in and sticking the drill string or forming large caverns.
- Provides a strong upper foundation to use high-density drilling fluid to continue drilling deeper.
- Isolates different zones, that may have different pressures or fluids - known as zonal isolation, in the drilled formations from one another.
- Seals off high pressure zones from the surface, avoiding potential for a blowout

- Prevents fluid loss into or contamination of production zones.
- Provides a smooth internal bore for installing production equipment.

A slightly different metal string, called production tubing, is often used without cement in the smallest casing of a well completion to contain production fluids and convey them to the surface from an underground reservoir.

Design

In the planning stages of a well a drilling engineer, usually with input from geologists and others, will pick strategic depths at which the hole will need to be cased in order for drilling to reach the desired total depth. This decision is often based on subsurface data such as formation pressures, strengths, and makeup, and is balanced against the cost objectives and desired drilling strategy. With the casing set depths determined, hole sizes and casing sizes must follow. The hole drilled for each casing string must be large enough to easily fit the casing inside it, allowing room for cement between the outside of the casing and the hole. Also, the inside diametre of the first casing string must be large enough to fit the second bit that will continue drilling. Thus, each casing string will have a subsequently smaller diametre. The inside diametre of the final casing string (or penultimate one in some instances of a liner completion) must accommodate the production tubing and associated hardware such as packers, gas lift mandrels and subsurface safety valves. Casing design for each size is done by calculating the worst condition that may be faced during drilling and production. Mechanical properties of designed pipes such as collapse resistance, burst pressure, and tensile strength must be sufficient for the worst conditions. Casing strings are supported by casing hangers that are set in the wellhead, which later will be topped with the Christmas tree. The wellhead usually is installed on top of the first casing string after it has been cemented in place.

Intervals

Typically, a well contains multiple intervals of casing successively placed within the previous casing run. The following casing interval is typically used in an oil or gas well:

- Conductor casing
- Surface casing
- Intermediate casing (optional)
- Production casing
- Production liner

The conductor casing serves as a support during drilling operations, to flowback returns during drilling and cementing of the surface casing, and to prevent collapse of the loose soil near the surface. It can normally vary from sizes such as 18" to 30".

The purpose of surface casing is to isolate freshwater zones so that they are not contaminated during drilling and completion. Surface casing is the most strictly regulated due to these environmental concerns, which can include regulation of casing depth and cement quality. A typical size of surface casing is 13\! inches.

Intermediate casing may be necessary on longer drilling intervals where necessary drilling mud weight to prevent blowouts may cause a hydrostatic pressure that can fracture deeper formations. Casing placement is selected so that the hydrostatic pressure of the drilling fluid remains between formation pore and fracture pressures.

The final interval is production casing. As with the casing intervals described above, the production casing string extends to the surface where it is hung off. As the smallest casing, it will form the outer boundary of the 'A' annulus, which may involve it being used for gas lift and well kills. A typical size is 9]! inches.

In order to reduce cost, a liner may be used which extends just above the shoe (bottom) of the previous casing interval and hung off downhole rather than at the surface. It may typically be 7", although many liners match the diametre of the production tubing.

Cementing

Cementing is performed by circulating a cement slurry through the inside of the casing and out into the annulus through the casing shoe at the bottom of the casing string. In order to precisely place the cement slurry at a required interval on the outside of the casing, a plug is pumped with a displacement fluid behind the cement slurry column, which "bumps" in the casing shoe and prevents further flow of fluid through the shoe. This bump can be seen at surface as a pressure spike at the cement pump. To prevent the cement from flowing back into the inside of the casing, a float collar above the casing shoe acts as a check valve and prevents fluid from flowing up through the shoe from the annulus.

Chapter 5

Petroleum Drilling Industry

Oil and natural gas drilling rigs can be used not only to identify geologic reservoirs but also to create holes that allow the extraction of oil or natural gas from those reservoirs. Primarily in onshore oil and gas fields once a well has been drilled, the drilling rig will be moved off of the well and a service rig (a smaller rig) that is purpose-built for completions will be moved on to the well to get the well on line. This frees up the drilling rig to drill another hole and streamlines the operation as well as allowing for specialisation of certain services, i.e., completions vs. drilling.

Water Well Drilling

New portable drillcat technology uses smaller portable trailer mounted rigs with shorter 10 foot (3.0 m) drill pipe. DIY users and missionary groups use these to drill water wells as they can be operated by 1 or 2 people with a minimal skill level. The shorter drill pipe also allows a much smaller mast, which gives a smaller and lighter rig which is cheaper to ship overseas and can fit in a standard 20 foot (6.1 m) shipping container. Portable trailer mounted drilling rigs have drill ratings from 300 to 800 feet (91 to 240 m) depending on mud pump flow and pressure ratings.

Other, heavier, truck rigs are more complicated, thus requiring more skill to run. They're also more difficult to handle safely due to the longer 20 to 30 foot (6.1 to 9.1 m) drill pipe. Large truck rigs also require a much higher over head clearance to operate. Large truck drills can use over 150 US gallons (570 L) or more of fuel per day, while the smaller portable drills use a mere 5 to 20 US gallons (19 to 76 L) of fuel per day. This makes smaller, more portable rigs preferable in remote or hard-to-reach places, and are more cost effective in this new era of high fuel prices.

History

Until internal combustion engines were developed in the late 19th century, the main method for drilling rock was muscle power of man or animal. Rods were turned by hand, using clamps attached to the rod. The rope and drop method invented in Zigong, China used a steel rod or piston raised and dropped vertically via a rope. Mechanised versions of this persisted until about 1970, using a cam to rapidly raise and drop what, by then, was a steel cable.

In the 1970s, outside of the oil and gas industry, roller bits using mud circulation were replaced by the first pneumatic reciprocating piston Reverse Circulation (RC) drills, and became essentially obsolete for most shallow drilling, and are now only used in certain situations where rocks preclude other methods. RC drilling proved much faster and more efficient, and continues to improve with better metallurgy, deriving harder, more durable bits, and compressors delivering higher air pressures at higher volumes, enabling deeper and faster penetration. Diamond drilling has remained essentially unchanged since its inception.

Mobile Drilling Rigs

In early oil exploration, drilling rigs were semi-permanent in nature and the derricks were often built on site and left in place after the completion of the well. In more recent times drilling rigs are expensive custom-built machines that can be moved from well to well. Some light duty drilling rigs are like a mobile crane and are more usually used to drill water wells. Larger land rigs must be broken apart into sections and loads to move to a new place, a process which can often take weeks. Small mobile drilling rigs are also used to drill or bore piles. Rigs can range from 100 ton continuous flight auger (CFA) rigs to small air powered rigs used to drill holes in quarries, etc. These rigs use the same technology and equipment as the oil drilling rigs, just on a smaller scale.

The drilling mechanisms outlined below differ mechanically in terms of the machinery used, but also in terms of the method by which drill cuttings are removed from the cutting face of the drill and returned to surface.

Drilling Rig Classification: There are many types and designs of drilling rigs, with many drilling rigs capable of switching or combining different drilling technologies as needed. Drilling rigs can be described using any of the following attributes:

By Power Used

- Mechanical — the rig uses torque converters, clutches, and transmissions powered by its own engines, often diesel
- Electric — the major items of machinery are driven by electric motors, usually with power generated on-site using internal combustion engines
- Hydraulic — the rig primarily uses hydraulic power
- Pneumatic — the rig is primarily powered by pressurized air
- Steam — the rig uses steam-powered engines and pumps (obsolete after middle of 20th Century).

By Pipe Used

- Cable — a cable is used to raise and drop the drill bit
- Conventional — uses metal or plastic drill pipe of varying types
- Coil tubing — uses a giant coil of tube and a downhole drilling motor.

By Height

(All rigs drill with only a single pipe. Rigs are differentiated by how many connected pipe they are able to "stand" in the derrick when needing to temporarily remove the drill pipe from the hole. Typically this is done when changing a drill bit or when "logging" the well.)

- Single — can pull only single drill pipes. The presence or absence of vertical pipe racking "fingers" varies from rig to rig.
- Double — can hold a stand of pipe in the derrick consisting of two connected drill pipes, called a "double stand".
- Triple — can hold a stand of pipe in the derrick consisting of three connected drill pipes, called a "triple stand".

By Method of Rotation or Drilling Method

- No-rotation includes direct push rigs and most service rigs.

Rotary Table (Drilling Rig)

A rotary table is a mechanical device on a drilling rig that provides clockwise (as viewed from above) rotational force to the drill string to facilitate the process of drilling a borehole.

The *kelly drive* #19, is inserted through the centre of the rotary table and kelly bushings, and has free vertical (up & down) movement to allow downward force to be applied to the drill string, while the rotary table rotates it. (Note: Force is not actually applied from the top (as to push) but rather the weight is at the bottom of the drill string like a pendulum on a string.)

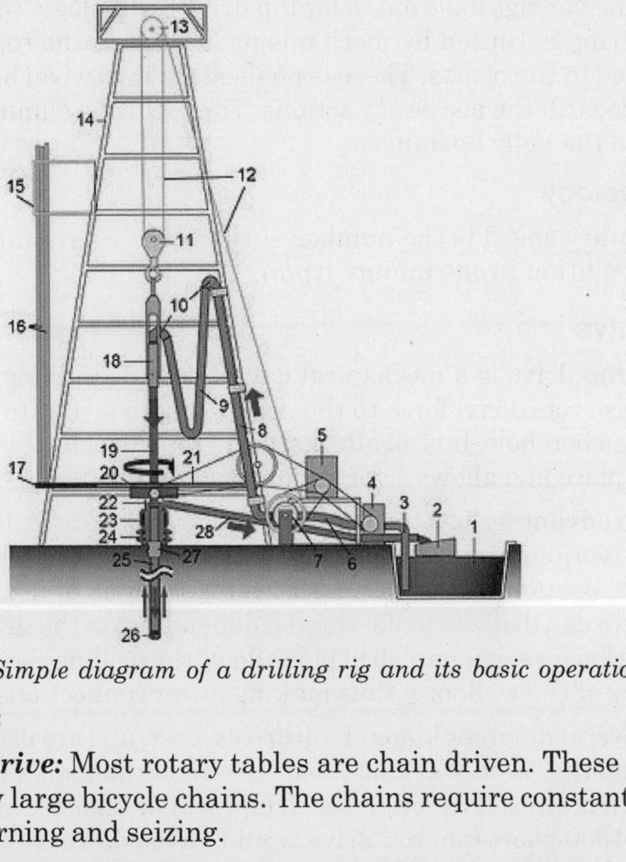

Figure: *Simple diagram of a drilling rig and its basic operation*

Components

Chain Drive: Most rotary tables are chain driven. These chains resemble very large bicycle chains. The chains require constant oiling to prevent burning and seizing.

Rotary Locks: Virtually all rotary tables are equipped with a rotary lock. Engaging the lock can either prevent the rotary from turning in one particular direction, or from turning at all. This is commonly used by crews in lieu of using a second pair of tongs to makeup or break out pipe.

Bushings: The rotary bushings are located at the centre of the rotary table. These can generally be removed in two separate pieces to facilitate large items, i.e. drill bits, to pass through the rotary table. The large gap in the centre of the rotary bushings is referred to as the "bowl" due to its appearance. The bowl is where the slips are set to hold up the drill string during connections and pipe trips as well as the point the drill string passes through the floor into the wellbore. The rotary bushings connect to the kelly bushings to actually induce the spin required for drilling.

Alternatives

Most recently manufactured rigs no longer feature rotary drives. These newer rigs have opted for top drive technology. In top drive, the drill string is turned by mechanisms located in the top drive that is attached to the blocks. There is no need for the swivel because the top drive does all the necessary actions. The top drive eliminates the kelly bar and the kelly bushings.

Terminology

Rotary speed is the number of times the rotary table makes one full revolution in one minute (rpm).

Top Drive

A top drive is a mechanical device on a drilling rig that provides clockwise rotational force to the drill string to facilitate the process of drilling a borehole. It is an alternative to rotary table. It is located at the swivel place and allows a vertical movement up and down the derrick.

An advantage of a top drive is that it allows the drilling rig to drill longer sections of a stand of drill pipe. A rotary table type rig can only drill 30'-60' (double-triple respectively) sections of drill pipe while a top drive can drill 60' to 90' stands, depending on the drilling rig type. Having longer sections of drill pipe allows the drilling rigs to drill deeper sections of the wellbore, thus making fewer connections of drill pipe.

Several different kinds of top drives exist, and are usually classified based on the "Safe Working Load" (SWL) of the tool, and the size and type of motor used to rotate the drillpipe. For offshore and heavy duty use, a 1000 short ton, top drive would be used, where a smaller land rig may only require a 500. Motors are available in all sizes, and come in Hydraulically powered, AC, or DC motors.

A top drive is a hydraulic or electric motor suspended in the derrick (mast) of a drilling rig, which rotates the drill string and bit and is used in the actual process of drilling the well. Using a top drive eliminates the need for the traditional kelly and rotary table and reduces the amount of manual labour and associated hazards that have traditionally accompanied this task. For drilling contractors and producers, utilising top drive technology translates into increased efficiency in its operations. Top drive ratings are based on the safe static hook load, this rating relates to the main thrust bearing (usually the weakest link). Top drive ratings of 500 short tons typically have an API rating lower than 500T, due to required calculations of rotation with a percentage of rated load.

Down-the-Hole Drill

Down-The-Hole Drill is called DTH in most drilling terms. The down-the-hole drill is basically a mini jack hammer that screws on the bottom of a drill string. The fast hammer action breaks hard rock into small flakes and dust and is blown clear by the air exhaust from the DTH hammer. The DTH hammer is one of the fastest ways to drill hard rock. Now smaller portable drillcat drilling rigs with DTH hammers can drill as fast as much larger truck rigs with this newer technology.

History

A pneumatic tool is first thought to have been used for rock drilling in 1844. Mr Burton invented a "wind hammer" for boring holes in mines.

Origin of the Name

DTH is short for "down-the-hole". Since the DTH method was originally developed to drill large-diametre holes downwards in surface-drilling applications, its name originated from the fact that the percussion mechanism followed the bit down into the hole. Applications were later found for the DTH method underground, where the direction of drilling is generally upwards instead of downwards.

Technical Details

In *DTH* drilling, the percussion mechanism – commonly called the hammer – is located directly behind the drill bit. The drill pipes transmit the necessary feed force and rotation to hammer and bit plus compressed air or fluids for the hammer and flushing of cuttings. The drill pipes are added to the drill string successively behind the hammer as the hole gets deeper. The piston strikes the impact surface of the bit directly, while the hammer casing gives straight and stable guidance of the drill bit. This means that the impact energy does not have to pass through any joints at all. The impact energy therefore is not lost in joints allowing for much deeper percussion drilling.

Uses

DTH drilling is used in the construction industry to produce piles into rock, also water wells, and drilling bores for geothermal ground source heat pumps.

Drill Types

There are a variety of drill mechanisms which can be used to sink a borehole into the ground. Each has its advantages and disadvantages, in terms of the depth to which it can drill, the type of sample returned, the costs involved and penetration rates achieved. There are two basic types of drills: drills which produce rock chips, and drills which produce core samples.

Auger Drilling

Auger drilling is done with a helical screw which is driven into the ground with rotation; the earth is lifted up the borehole by the blade of the screw. Hollow stem auger drilling is used for softer ground such as swamps where the hole will not stay open by itself for environmental drilling, geotechnical drilling, soil engineering and geochemistry reconnaissance work in exploration for mineral deposits. Solid flight augers/bucket augers are used in harder ground construction drilling. In some cases, mine shafts are dug with auger drills. Small augers can be mounted on the back of a utility truck, with large augers used for sinking piles for bridge foundations.

Auger drilling is restricted to generally soft unconsolidated material or weak weathered rock. It is cheap and fast.

Figure: *Cable tool water well drilling rig in Kimball, West Virginia. These slow rigs have mostly been replaced by rotary drilling rigs in the U.S.*

Percussion Rotary Air Blast Drilling (RAB)

RAB drilling is used most frequently in the mineral exploration industry. (This tool is also known as a Down-the-hole drill.) The drill uses a pneumatic reciprocating piston-driven “hammer” to energetically drive a heavy drill bit into the rock. The drill bit is hollow, solid steel and has ~20 mm thick tungsten rods protruding from the steel matrix as "buttons". The tungsten buttons are the cutting face of the bit.

The cuttings are blown up the outside of the rods and collected at surface. Air or a combination of air and foam lift the cuttings.

RAB drilling is used primarily for mineral exploration, water bore drilling and blast-hole drilling in mines, as well as for other applications such as engineering, etc. RAB produces lower quality samples because the cuttings are blown up the outside of the rods and can be contaminated from contact with other rocks. RAB drilling at extreme depth, if it encounters water, may rapidly clog the outside of the hole with debris, precluding removal of drill cuttings from the hole. This can be counteracted, however, with the use of "stabilisers" also known as "reamers", which are large cylindrical pieces of steel attached to the drill string, and made to perfectly fit the size of the hole being drilled.

The use of high-powered air compressors, which push 900-1150 cfm of air at 300-350 psi down the hole also ensures drilling of a deeper hole up to ~1250 m due to higher air pressure which pushes all rock cuttings and any water to the surface. This, of course, is all dependent on the density and weight of the rock being drilled, and on how worn the drill bit is.

Air Core Drilling

Air core drilling and related methods use hardened steel or tungsten blades to bore a hole into unconsolidated ground. The drill bit has three blades arranged around the bit head, which cut the unconsolidated ground. The rods are hollow and contain an inner tube which sits inside the hollow outer rod barrel. The drill cuttings are removed by injection of compressed air into the hole via the annular area between the innertube and the drill rod. The cuttings are then blown back to surface up the inner tube where they pass through the sample separating system and are collected if needed. Drilling continues with the addition of rods to the top of the drill string. Air core drilling can occasionally produce small chunks of cored rock.

This method of drilling is used to drill the weathered regolith, as the drill rig and steel or tungsten blades cannot penetrate fresh rock. Where possible, air core drilling is preferred over RAB drilling as it provides a more representative sample. Air core drilling can achieve depths approaching 300 metres in good conditions. As the cuttings are removed inside the rods and are less prone to contamination compared to conventional drilling where the cuttings pass to the surface via outside return between the outside of the drill rod and the walls of the hole. This method is more costly and slower than RAB.

Cable Tool Drilling

Cable tool rigs are a traditional way of drilling water wells. The majority of large diametre water supply wells, especially deep wells completed in bedrock aquifers, were completed using this drilling method. Although this drilling method has largely been supplanted in recent years by other, faster drilling techniques, it is still the most practicable drilling method for large diametre, deep bedrock wells, and in widespread use for small rural water supply wells. The impact of the drill bit fractures the rock and in many shale rock situations increases the water flow into a well over rotary.

Also known as ballistic well drilling and sometimes called "spudders", these rigs raise and drop a drill string with a heavy carbide tipped drilling bit that chisels through the rock by finely pulverizing the subsurface materials. The drill string is composed of the upper drill rods, a set of "jars" (inter-locking "sliders" that help transmit additional energy to the drill bit and assist in removing the bit if it is stuck) and the drill bit. During the drilling process, the drill string is periodically removed from the borehole and a bailer is lowered to collect the drill cuttings (rock fragments, soil, etc.). The bailer is a bucket-like tool with a trapdoor in the base. If the borehole is dry, water is added so that the drill cuttings will flow into the bailer. When lifted, the trapdoor closes and the cuttings are then raised and removed. Since the drill string must be raised and lowered to advance the boring, the casing (larger diametre outer piping) is typically used to hold back upper soil materials and stabilise the borehole.

Cable tool rigs are simpler and cheaper than similarly sized rotary rigs, although loud and very slow to operate. The world record cable tool well was drilled in New York to a depth of almost 12,000 feet (3,700 m). The common Bucyrus Erie 22 can drill down to about 1,100 feet (340 m). Since cable tool drilling does not use air to eject the drilling chips like a rotary, instead using a cable strung bailer, technically there is no limitation on depth.

Cable tool rigs now are nearly obsolete in the United States. They are mostly used in Africa or Third-World countries. Being slow, cable tool rig drilling means increased wages for drillers. In the United States drilling wages would average around US$200 per day per man, while in Africa it is only US$6 per day per man, so a slow drilling machine can still be used in undeveloped countries with depressed wages. A cable tool rig can drill 25 feet (7.6 m) to 60 feet (18 m) of hard rock a day. A newer rotary drillcat top head rig equipped with down-the-hole (DTH) hammer can drill 500 feet (150 m) or more per day, depending on size and formation hardness.

Reverse Circulation (RC) Drilling

RC drilling is similar to air core drilling, in that the drill cuttings are returned to surface inside the rods. The drilling mechanism is a pneumatic reciprocating piston known as a "hammer" driving a tungsten-steel drill bit. RC drilling utilises much larger rigs and machinery and depths of up to 500 metres are routinely achieved. RC drilling ideally produces dry rock chips, as large air compressors dry the rock out ahead of the advancing drill bit. RC drilling is slower and costlier but achieves better penetration than RAB or air core drilling; it is cheaper than diamond coring and is thus preferred for most mineral exploration work.

Reverse circulation is achieved by blowing air down the rods, the differential pressure creating air lift of the water and cuttings up the "inner tube", which is inside each rod. It reaches the "bell" at the top of the hole, then moves through a sample hose which is attached to the top of the "cyclone". The drill cuttings travel around the inside of the cyclone until they fall through an opening at the bottom and are collected in a sample bag.

The most commonly used RC drill bits are 5-8 inches (13–20 cm) in diametre and have round metal 'buttons' that protrude from the bit, which are required to drill through shale and abrasive rock. As the buttons wear down, drilling becomes slower and the rod string can potentially become bogged in the hole. This is a problem as trying to recover the rods may take hours and in some cases weeks. The rods and drill bits themselves are very expensive, often resulting in great cost to drilling companies when equipment is lost down the bore hole. Most companies will regularly re-grind the buttons on their drill bits in order to prevent this, and to speed up progress. Usually, when something is lost (breaks off) in the hole, it is not the drill string, but rather from the bit, hammer, or stabiliser to the bottom of the drill string (bit). This is usually caused by a blunt bit getting stuck in fresh rock, over-stressed metal, or a fresh drill bit getting stuck in a part of the hole that is too small, owing to having used a bit that has worn to smaller than the desired hole diametre.

Although RC drilling is air-powered, water is also used, to reduce dust, keep the drill bit cool, and assist in pushing cutting back upwards, but also when "collaring" a new hole. A mud called "Liqui-Pol" is mixed with water and pumped into the rod string, down the hole. This helps to bring up the sample to the surface by making the sand stick together. Occasionally, "Super-Foam" (a.k.a. "Quik-Foam") is also used, to bring all the very fine cuttings to the surface, and to clean the hole. When the drill reaches hard rock, a "collar" is put down the hole around the rods, which is normally PVC piping. Occasionally the collar may be made from metal casing. Collaring a hole is needed to stop the walls from caving in and bogging the rod string at the top of the hole. Collars may be up to 60 metres deep, depending on the ground, although if drilling through hard rock a collar may not be necessary.

Reverse circulation rig setups usually consist of a support vehicle, an auxiliary vehicle, as well as the rig itself. The support vehicle, normally a truck, holds diesel and water tanks for resupplying the rig. It also holds other supplies needed for maintenance on the rig. The auxiliary is a vehicle, carrying an auxiliary engine and a booster engine.

These engines are connected to the rig by high pressure air hoses. Although RC rigs have their own booster and compressor to generate air pressure, extra power is needed which usually isn't supplied by the rig due to lack of space for these large engines. Instead, the engines are mounted on the auxiliary vehicle. Compressors on an RC rig have an output of around 1000 cfm at 500 psi (500 $L \cdot s^{-1}$ at 3.4 MPa). Alternatively, stand-alone air compressors which have an output of 900-1150cfm at 300-350 psi each are used in sets of 2, 3, or 4, which are all routed to the rig through a multi-valve manifold.

Diamond Core Drilling

Diamond core drilling (exploration diamond drilling) utilises an annular diamond-impregnated drill bit attached to the end of hollow drill rods to cut a cylindrical core of solid rock. The diamonds used are fine to microfine industrial grade diamonds. They are set within a matrix of varying hardness, from brass to high-grade steel. Matrix hardness, diamond size and dosing can be varied according to the rock which must be cut. Holes within the bit allow water to be delivered to the cutting face. This provides three essential functions — lubrication, cooling, and removal of drill cuttings from the hole.

Diamond drilling is much slower than reverse circulation (RC) drilling due to the hardness of the ground being drilled. Drilling of 1200 to 1800 metres is common and at these depths, ground is mainly hard rock. Diamond rigs need to drill slowly to lengthen the life of drill bits and rods, which are very expensive.

Core samples are retrieved via the use of a core tube, a hollow tube placed inside the rod string and pumped with water until it locks into the core barrel. As the core is drilled, the core barrel slides over the core as it is cut. An "overshot" attached to the end of the winch cable is lowered inside the rod string and locks on to the backend (aka head assembly), located on the top end of the core barrel. The winch is retracted, pulling the core tube to the surface. The core does not drop out of the inside of the core tube when lifted because either a split ring core lifter or basket retainer allow the core to move into, but not back out of the tube.

Once the core tube is removed from the hole, the core sample is then removed from the core tube and catalogued. The Driller's assistant unscrews the backend off the core tube using tube wrenches, then each part of the tube is taken and the core is shaken out into core trays. The core is washed, measured and broken into smaller pieces using a hammer or sawn through to make it fit into the sample trays. Once catalogued, the core trays are retrieved by geologists who then analyse the core and determine if the drill site is a good location to expand future mining operations.

Diamond rigs can also be part of a multi-combination rig. Multi-combination rigs are a dual setup rig capable of operating in either a reverse circulation (RC) and diamond drilling role (though not at the same time). This is a common scenario where exploration drilling is being performed in a very isolated location. The rig is first set up to drill as an RC rig and once the desired metres are drilled, the rig is set up for diamond drilling. This way the deeper metres of the hole can be drilled without moving the rig and waiting for a diamond rig to set up on the pad.

Direct Push Rigs

Direct push technology includes several types of drilling rigs and drilling equipment which advances a drill string by pushing or hammering without rotating the drill string. While this does not meet the proper definition of drilling, it does achieve the same result — a borehole. Direct push rigs include both cone penetration testing (CPT) rigs and direct push sampling rigs such as a PowerProbe or Geoprobe. Direct push rigs typically are limited to drilling in unconsolidated soil materials and very soft rock.

CPT rigs advance specialised testing equipment (such as electronic cones), and soil samplers using large hydraulic rams. Most CPT rigs are heavily ballasted (20 metric tons is typical) as a counter force against the pushing force of the hydraulic rams which are often rated up to 20 kN. Alternatively, small, light CPT rigs and offshore CPT rigs will use anchors such as screwed-in ground anchors to create the reactive force. In ideal conditions, CPT rigs can achieve production rates of up to 250–300 metres per day.

Direct push drilling rigs use hydraulic cylinders and a hydraulic hammer in advancing a hollow core sampler to gather soil and groundwater samples. The speed and depth of penetration is largely dependent on the soil type, the size of the sampler, and the weight and power the rig. Direct push techniques are generally limited to shallow soil sample recovery in unconsolidated soil materials. The advantage of direct push technology is that in the right soil type it can produce a large number of high quality samples quickly and cheaply, generally from 50 to 75 metres per day. Rather than hammering, direct push can also be combined with sonic (vibratory) methods to increase drill efficiency.

Hydraulic Rotary Drilling

Oil well drilling utilises tri-cone roller, carbide embedded, fixed-cutter diamond, or diamond-impregnated drill bits to wear away at the cutting face. This is preferred because there is no need to return intact samples to surface for assay as the objective is to reach a formation containing oil or natural gas. Sizable machinery is used, enabling depths of several kilometres to be penetrated. Rotating hollow drill pipes carry down bentonite and barite infused drilling muds to lubricate, cool, and clean the drilling bit, control downhole pressures, stabilise the wall of the borehole and remove drill cuttings. The mud travels back to the surface around the outside of the drill pipe, called the annulus. Examining rock chips extracted from the mud is known as mud logging. Another form of well logging is electronic and is frequently employed to evaluate the existence of possible oil and gas deposits in the borehole. This can take place while the well is being drilled, using Measurement While Drilling tools, or after drilling, by lowering measurement tools into the newly drilled hole.

The rotary system of drilling was in general use in Texas in the early 1900s. It is a modification of one invented by Fauvelle in 1845, and used in the early years of the oil industry in some of the oil-producing countries in Europe. Originally pressurized water was used instead of mud, and was almost useless in hard rock before the diamond cutting bit. The main breakthrough for rotary drilling came in 1901, when Anthony Francis Lucas combined the use of a steam-driven rig and of mud instead of water in the Spindletop discovery well.

The drilling and production of oil and gas can pose a safety risk and a hazard to the environment from the ignition of the entrained gas causing dangerous fires and also from the risk of oil leakage polluting water, land and groundwater. For these reasons, redundant safety systems and highly trained personnel are required by law in all countries with significant production.

Sonic (Vibratory) Drilling

A sonic drill head works by sending high frequency resonant vibrations down the drill string to the drill bit, while the operator controls these frequencies to suit the specific conditions of the soil/rock geology. Vibrations may also be generated within the drill head. The frequency is generally between 50 and 120 hertz (cycles per second) and can be varied by the operator.

Resonance magnifies the amplitude of the drill bit, which fluidizes the soil particles at the bit face, allowing for fast and easy penetration through most geological formations. An internal spring system isolates these vibrational forces from the rest of the drill rig.

Limits of the Technology

Figure: *An oil rig*

Drill technology has advanced steadily since the 19th century. However, there are several basic limiting factors which will determine the depth to which a bore hole can be sunk. All holes must maintain outer diametre; the diametre of the hole must remain wider than the diametre of the rods or the rods cannot turn in the hole and progress cannot continue. Friction caused by the drilling operation will tend to reduce the outside diametre of the drill bit. This applies to all drilling methods, except that in diamond core drilling the use of thinner rods and casing may permit the hole to continue. Casing is simply a hollow sheath which protects the hole against collapse during drilling, and is made of metal or PVC. Often diamond holes will start off at a large diametre and when outside diametre is lost, thinner rods put down inside casing to continue, until finally the hole becomes too narrow. Alternatively, the hole can be reamed; this is the usual practice in oil well drilling where the hole size is maintained down to the next casing point.

For percussion techniques, the main limitation is air pressure. Air must be delivered to the piston at sufficient pressure to activate the reciprocating action, and in turn drive the head into the rock with sufficient strength to fracture and pulverise it. With depth, volume is added to the in-rod string, requiring larger compressors to achieve operational pressures. Secondly, groundwater is ubiquitous, and increases in pressure with depth in the ground. The air inside the rod string must be pressurised enough to overcome this water pressure at the bit face. Then, the air must be able to carry the rock fragments to surface. This is why depths in excess of 500 m for reverse circulation drilling are rarely achieved, because the cost is prohibitive and approaches the threshold at which diamond core drilling is more economic.

Diamond drilling can routinely achieve depths in excess of 1200 m. In cases where money is no issue, extreme depths have been achieved, because there is no requirement to overcome water pressure. However, circulation must be maintained to return the drill cuttings to surface, and more importantly to maintain cooling and lubrication of the cutting surface. Without sufficient lubrication and cooling, the matrix of the drill bit will soften. While diamond is the hardest substance known, at 10 on the Mohs hardness scale, it must remain firmly in the matrix to achieve cutting. Weight on bit, the force exerted on the cutting face of the bit by the drill rods in the hole above the bit, must also be monitored.

A unique drilling operation in deep ocean water was named Project Mohole.

Research of New Drilling Technologies

Limits of the conventional contact drilling technologies caused strengthen of the research of new non-contact effective drilling technologies. There were several attempts to achieve sufficient results of the research which would negate disadvantages of current contact technology. The best known are technologies based on the utilisation of water jet, chemical plasma, hydrothermal spallation or laser. The research teams round the world have been developing these technologies for the long time.

Nowadays, utilisation of high energetic electrical plasma shows very promising in deep drilling applications. This approach has potential to replace conventional drilling technologies because of several advantages.

Causes of Deviation

Most drill holes deviate slightly from their planned trajectory. This is because of the torque of the turning bit working against the cutting face, because of the flexibility of the steel rods and especially the screw joints, because of reaction to foliation and structure within the rock, and because of refraction as the bit moves into different rock layers of varying resistance. Additionally, inclined holes will tend to deviate upwards because the drill rods will lie against the bottom of the bore, causing the drill bit to be slightly inclined from true. It is because of deviation that drill holes must be surveyed if deviation will impact the usefulness of the information returned. Sometimes the surface location can be offset laterally to take advantage of the expected deviation tendency, so the bottom of the hole will end up near the desired location.

Rig Equipment

Drilling rigs typically include at least some of the following items:

- Blowout preventers: (BOPs)

The equipment associated with a rig is to some extent dependent on the type of rig but (#23 & #24) are devices installed at the wellhead to prevent fluids and gases from unintentionally escaping from the borehole. #23 is the annular (often referred to as the "Hydril", which is one manufacturer) and #24 is the pipe rams and blind rams. In the place of #24 Variable bore rams or VBR's can be used, they offer the same pressure and sealing capacity found in standard pipe rams, while offering the versatility of sealing on various sizes of drill pipe, production tubing and casing without changing standard pipe rams.

- Centrifuge: an industrial version of the device that separates fine silt and sand from the drilling fluid.
- Solids control: solids control equipments for preparing drilling mud for the drilling rig.

- Chain tongs: wrench with a section of chain, that wraps around whatever is being tightened or loosened. Similar to a pipe wrench.
- Degasser: a device that separates air and/or gas from the drilling fluid.
- Desander / desilter: contains a set of hydrocyclones that separate sand and silt from the drilling fluid.
- Drawworks: (#7) is the mechanical section that contains the spool, whose main function is to reel in/out the drill line to raise/ lower the travelling block (#11).
- Drill bit: (#26) device attached to the end of the drill string that breaks apart the rock being drilled. It contains jets through which the drilling fluid exits.
- Drill pipe: (#16) joints of hollow tubing used to connect the surface equipment to the bottom hole assembly (BHA) and acts as a conduit for the drilling fluid. In the diagram, these are "stands" of drill pipe which are 2 or 3 joints of drill pipe connected together and "stood" in the derrick vertically, usually to save time while tripping pipe.
- Elevators: a gripping device that is used to latch to the drill pipe or casing to facilitate the lowering or lifting (of pipe or casing) into or out of the borehole.
- Mud motor: a hydraulically powered device positioned just above the drill bit used to spin the bit independently from the rest of the drill string.
- Mud pump: (#4) reciprocal type of pump used to circulate drilling fluid through the system.
- Mud tanks: (#1) often called mud pits, provides a reserve store of drilling fluid until it is required down the wellbore.
- Rotary table: (#20) rotates the drill string along with the attached tools and bit.
- Shale shaker: (#2) separates drill cuttings from the drilling fluid before it is pumped back down the borehole.

List of Components of Oil Drilling Rigs

Offshore drilling rigs have similar elements, but are configured with a number of different drilling systems to suit drilling in the marine environment. The equipment associated with a rig is to some extent dependent on the type of rig but typically includes at least some of the items listed below.

Mud Tank

A mud tank is an open-top container, typically made of steel, to store drilling fluid on a drilling rig. They are also called *mud pits*, because they used to be nothing more than pits dug out of the earth.

The tanks are generally open-top and have walkways on top to allow a worker to traverse and inspect the level of fluid in the tanks. The walkways also allow access to other equipment mounted on top. Recently, offshore drilling rigs have closed-in tanks for safety. The mud tank plays a critical role in mechanically removing destructive solids and sediment from costly land and offshore drilling systems.

A drilling rig normally has two tanks. A tank is sectioned off into compartments. A compartment may include a settling tank, sometimes called a sand trap, to allow sand and other solids in the drilling fluid to precipitate before it flows into the next compartment. Other compartments may have agitators, which are large fan blades stirring the fluid to prevent its contents from precipitating.

Mud Tanks for Solids Control

Mud tanks are the base of solids control equipment. Mud tanks for solids control systems are divided into Square tanks and Cone-shaped tanks according to the shape of the tank bottom. The body of the tank is made by welding the steel plate and section, using the smooth cone-shape structure or the corrugated structure. The mud tank surface and passages are made of slip resistant steel plate and expanded steel plate. The mud tanks are made of the side steel pipe, all of the structure can be folded without barrier and pegged reliably. The surface of the tank is equipped with a water pipeline for cleaning the surface and equipment on the tank, it uses soaked zinc processing for the expanded steel plate. The ladder is made of channel steel to take responsibility the body, the foot board is made of expanded steel plate. The two-sided guard rail are installed the safe suspension hook. The mud tank is designed the standard shanty to prevent the sand and the rain. The pipeline is installed in the tank to preserve the warm air heat.

Mud tank quantity depends on mud process capacity requirements. Especially for mud storage, suitable for mud pump, drilling rig, etc.

Shale Shakers

Shale shakers are devices that remove drill cuttings from the drilling fluid that is used for boring holes into the earth. Controlling the solids in drilling fluid is an important component of the cost of drilling, so research into improved shaker design is ongoing.

Shale Shaker Design

Shale shakers typically consist of large, flat sheets of wire mesh screens or sieves of various mesh sizes that shake or vibrate the drill cuttings, commonly shale, across and off of the screens as the drilling fluid (mud) flows through them and back into the drilling fluid system. This separates the solid drill cuttings from the fluid so that it can be recirculated back down the wellbore. In oilfield industry, linear motion shale shakers are widely used.

Shaker Screen Panels

Screen panels play a major role by removing particles larger than the mesh size. Screen selection is critical since shaking is the primary stage in the removal of solids. Improper screen selection can lead to de-blinding, loss of drilling fluids, premature pump part failures, overloading of other solids removal equipment within the mud systems, decreased service life, reduced flow rate capacity, and serious problems in the wellbore.

Causes of Screen Failure

The causes of premature screen failure are:

- Personnel improperly trained on handling, storage, maintenance and installation of deck rubbers and screens
- Careless storage of screen panels prior to use
- Screens damaged before use from improper handling during installation
- Shaker screen not match very well to shale shaker
- Deck rubbers improperly installed
- Deck rubbers dirty, worn or missing
- Dried cuttings or drilling fluid left on screen during shutdown of shaker
- Personnel walking on screens or using the shaker as a work table
- Tools being dropped on screens
- Extremely high mud weights or heavy solids loading
- Defective or improperly manufactured screens
- Use of high pressure wash guns to clean plugged screens

Screen Panel Designation

The American Petroleum Institute (API) Screen Designation is the customary identification for screen panels. This includes:

- API Number: the sieve equivalent as per API RP 13C
- Conductance: the ease with which a liquid can flow through the screen, with larger values representing higher volume handling
- Microns: a unit of length equal to one-thousandth of a millimetre
- Non-blanked area: an evaluation of the surface area available for liquid transmission through the screen.

Suction Line

A suction line is the pipe work linking the mud tanks/pits with the mud pumps. This may be gravity fed or charged by centrifugal pumps to provide additional volumetric efficiency to the mud pumps.

Mud Pump

A mud pump is a reciprocating piston/plunger device designed to circulate drilling fluid under high pressure (up to 7,500 psi (52,000 kPa)) down the drill string and back up the annulus.

Mud pumps come in a variety of sizes and configurations but for the typical petroleum drilling rig, the *triplex* (three piston/plunger) mud pump is the pump of choice. Duplex mud pumps (two piston/plungers) have generally been replaced by the triplex pump, but are still common in developing countries. Two later developments are is the hex pump with six vertical pistons/plungers, and various quintuplex's with five horizontal piston/plungers. The advantages that these new pumps have over convention triplex pumps is a lower mud noise which assists with better MWD and LWD retrieval.

The "normal" mud pump consist of two main sub-assemblies, the fluid end and the power end.

The fluid end produces the pumping process with valves, pistons, and liners. Because these components are high-wear items, modern pumps are designed to allow quick replacement of these parts.

To reduce severe vibration caused by the pumping process, these pumps incorporate both a suction and discharge pulsation dampener. These are connected to the inlet and outlet of the fluid end. The power end converts the rotation of the drive shaft to the reciprocating motion of the pistons. In most cases a crosshead crank gear is used for this.

Parts of mud pump: 1.housing itself. 2.liner with packing. 3.cover plus packing. 4.piston and piston rod. 5.suction valve and discharge valve with their seats. 6.stuffing box (only in double-acting pumps). 7.gland (only in double-acting pumps).

Draw-Works

A draw-works is the primary hoisting machinery that is a component of a rotary drilling rig. Its main function is to provide a means of raising and lowering the travelling blocks. The wire-rope drilling line winds on the drawworks drum and extends to the crown block and travelling blocks, allowing the drill string to be moved up and down as the drum turns. The segment of drilling line from the draw-works to the crown block is called the "fast line". The drilling line then enters the sheaves of the crown block and is makes several passes between the crown block and travelling block pulleys for mechanical advantage. The line then exits the last sheave on the crown block and is fastened to a derrick leg on the other side of the rig floor. This section of drilling line is called the "dead line".

A modern draw-works consists of five main parts: the drum, the motor(s), the reduction gear, the brake, and the auxiliary brake. The motors can be AC or DC-motors, or the draw-works may be connected directly to diesel engines using metal chain-like belts. The number of gears could be one, two or three speed combinations. The main brake, usually operated manually by a long handle, may be friction band brake, a disc brake or a modified clutch. It serves as a parking brake when no motion is desired. The auxiliary brake is connected to the drum, and absorbs the energy released as heavy loads are lowered. This brake may use eddy current rotors or water-turbine-like apparatus to convert to heat the kinetic energy of a downward-moving load being stopped. Power catheads (winches) located on each side provide the means of actuating the tongs used to couple and uncouple threaded pipe members. Outboard catheads can be used manually with ropes for various small hoisting jobs around the rig.

The drawworks often has a pulley drive arrangement on the front side to provide turning power to the rotary table, although on many rigs the rotary table is independently powered.

Rig Standpipe

A rig standpipe is a solid metal pipe attached to the side of a drilling rig's derrick that is a part of its drilling mud system. It is used to conduct drilling fluid from the mud pumps to the kelly hose. Bull plugs, pressure transducers and valves are found on the rig standpipe.

Kelly Hose

A Kelly hose (also known as a *mud hose* or *rotary hose*) is a flexible, steel reinforced, high pressure hose that connects the standpipe to the kelly (or more specifically to the goose-neck on the swivel above the kelly) and allows free vertical movement of the kelly while facilitating the flow of drilling fluid through the system and down the drill string.Travelling block

A travelling block is the freely moving section of a block and tackle that contains a set of pulleys or sheaves through which the drill line (wire rope) is threaded or reeved and is opposite (and under) the crown block (the stationary section). The combination of the travelling block, crown block and wire rope drill line gives the ability to lift weights in the *hundreds of thousands* of pounds. On larger drilling rigs, when raising and lowering the derrick, line tensions over a million pounds are not unusual.

Travelling Block

A travelling block is the freely moving section of a block and tackle that contains a set of pulleys or sheaves through which the drill line (wire rope) is threaded or reeved and is opposite (and under) the crown block (the stationary section). The combination of the travelling block, crown block and wire rope drill line gives the ability to lift weights in the *hundreds of thousands* of pounds. On larger drilling rigs, when raising and lowering the derrick, line tensions over a million pounds are not unusual.

Drill Line

In a drilling rig, the drill line is a multi-thread, twisted *wire* rope that is threaded or reeved through the travelling block and crown block to facilitate the lowering and lifting of the drill string into and out of the wellbore. On larger diametre lines, tension strengths over a *million* pounds are possible. To make a connection is to add another segment of drill pipe onto the top the drill string. A segment is added by pulling the kelly above the rotary table, stopping the mud pump, hanging off the drill string in the rotary table, unscrewing the kelly from the drill pipe below, swinging the kelly over to permit connecting it to the top of the new segment (which had been placed in the mousehole), and then screwing this assembly into the top of the existing drill string. Mud circulation is resumed, and the drill string is lowered into the hole until the bit takes weight at the bottom of the hole. Drilling then resumes.

Crown Block

A crown block is the stationary section of a block and tackle that contains a set of pulleys or sheaves through which the drill line (wire rope) is threaded or reeved and is opposite and above the travelling block. The combination of the travelling block, crown block and wire rope drill line gives the ability to lift weights in the hundreds of thousands of pounds. On larger drilling rigs, when raising and lowering the derrick, line tensions over a million pounds are not unusual.

Derrick

A derrick is a lifting device composed of one tower, or guyed mast such as a pole which is hinged freely at the bottom. It is controlled by lines (usually four of them) powered by some means such as man-hauling or motors, so that the pole can move in all four directions. A line runs up it and over its top with a hook on the end, like with a crane. It is commonly used in docks and onboard ships. Some large derricks are mounted on dedicated vessels, and are often known as "floating derricks".

The device was named for its resemblance to a type of gallows from which a hangman's noose hangs. The derrick type of gallows in turn got its name from Thomas Derrick, an English executioner from the Elizabethan era.

Oil Derrick

Another kind of derrick is used over oil wells and other drilled holes. This is generally called an oil derrick and is a complex set of machines specifically designed for optimum efficiency, safety and low cost. This is used on some offshore oil and gas rigs.

The centrepiece is the archetypical derrick tower, used for lifting and positioning the drilling string and piping above the well bore, and containing the machinery for turning the drilling bit around in the hole. As the drill string goes deeper into the underlying soil or rock, new piping has to be added to the top of the drill to keep the connection between drill bit and turning machinery intact, to create a filler to keep the hole from caving in, and to create a conduit for the drilling mud. The drilling mud is used to cool the drilling bit and to blow rock debris clear from the drill bit and the bottom of the well. The piping joints sections—usually each about 10 metres (30 ft) long—have threaded ends, so they can be screwed together. The piping is hollow to allow for the mud to be pumped down into the drilling hole, where it flushes out at the drilling bit. The mud then proceeds upwards towards the surface on the outside of the piping, carrying the debris with it.

The derrick also controls the weight on the drilling bit, because the drill bit works at an optimum rate only when it is pushed with a precise degree of pressure relative to the rock beneath it. Too much weight can break the drilling bit, and not enough weight will prolong drilling time. At the start of drilling extra heavy piping collars are used to create enough weight to drill. Since the weight of the pipes above the drill bit will increase the pressure on it as it goes deeper, the derrick will apply less pressure as piping sections are added. It will eventually lift the nearly entire drill string-and-piping complex to prevent too much weight as the well goes deeper.

Patent Derrick Systems

Hallen Derrick: The boom is connected with the lower part of the mast which is shaped like a "Y" or a bipod and therefore it is a single swinging derrick. On the cross trees, two guys are fastened using swivel outriggers which are stayed vertically and horizontally. In order to maintain a good controlling angle between guys and derrick, the outriggers cannot pass the inboard parallel of the centreline. Looking at the illustration, one can easily see that the right outrigger stays in the centreline and the left outrigger has moved outboard. This derrick will lower or heave cargo as both guys are veered or hauled. Three winches, controlled by joystick, are necessary to operate the Hallen Derrick; two for the guys and one for the purchase. To avoid an over-topping or over-swinging limit-switches are used. However, the limits can be modified if a different working range or a special vertical stowage is required.

The safe working load (SWL) of the Hallen is between 10 and 80 tonnes. In a Hallen Universal derrick, which has no Hallen D-Frame, the halyard has an extended length since it runs through further blocks on the centreline. The Universal Hallen derrick, replacing the D-frame option, is a kind of traditional topping lift. The Hallen D-Frame is a steel bracket welded on the mast in the centreline. For an observer standing a beam, the frame has a "D"-shape. The D-Frame supersedes the outriggers and provides a good controlling angle on the guys.

The Hallen derrick has a good purpose for e.g. containers, logs, steel rail, sawn timber and heavy lifts and doesn't lend itself for small, general cargo. It keeps the deck clear of guy ropes and preventors. Only one winchman is needed and within a few minutes the Hallen is brought into use. It is less expensive than a crane. A disadvantage is the low working range of the Hallen Derrick, it is able to swing 75° from the centreline and can work against a list of up to 15°.

Velle Derrick

The Velle derrick is quite similar to the Hallen but without use of outriggers. On top of the boom is a T-shaped yoke assembled. Also here, the guys serve for topping and lowering the boom but they are fastened on the yoke with four short, steel-wire hanger-ropes. The ends of the topping and lowering ends of the halyard are secured to half-barrels on one winch.

In this way the boom moves in the same speed as the winch veers the topping end of the halyard and hauls the lowering end of the halyard, and vice versa. The slewing ends are also wound on to another half-barrel. For hoisting the cargo, there is a third winch to hoist to cargo on the yoke. Runners decrease swing and rotation of the cargo. A joystick duplex controller steers the Velle derrick.

Stülcken Derrick

The patent Stülcken derrick is used for very heavy cargo. It stems from the German shipyard HC Stülcken & Sohn which has been taken over later by neighbouring yard Blohm & Voss. This derrick can handle up to 300 tonnes. The Stülcken can be made ready in few minutes, which is a lot faster than a traditional heavy derrick, doesn't require lots of space and is operated by four winches. Between two v-shaped, unstayed Samson-posts is the Stülcken secured. This makes it possible to let the derrick swing through the posts to reach another hatch. For each post is a hoisting winch, a span winch and a lever that is run by one man only. Bearings, swivels, sheaves and the gooseneck can be unattended for up to four years and create only a friction of about 2%.

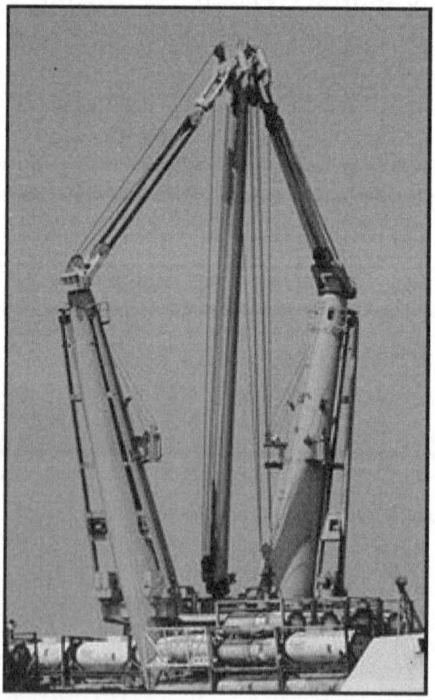

Figure: *Stülcken heavylift derrick*

The span tackles are independent and the halyard is endless. With the revolving suspension heads on the posts it takes ten minutes to swing all the way through. In the double-pendulum block type, half of the cargo tackle can be anchored to the base of the boom. In order to double the hook speed, the halyard passes through the purchases since one end is secured which reduces the SWL to its half. Typical dimensions of a 275 tonne Stülcken are: 25.5 m length, 0.97 m diametre, 1.5 m to 3.4 m diametre of posts, 18 m apart the posts (upper end) and 8.4 m apart the posts (lower end). The hook of a full-loaded 275 tonne Stülcken can move 2.3 m per minute. If only one purchase is secured and the derrick is loaded with 137 tonnes the hook gains velocity to 4.6 m per min.

Even more speed can be gained when the winch ratios are reduced to 100 tonnes (triple speed) and 68 tonnes (quadruple speed). Detaching the union table the double-pendulum block type of Stülcken is able to swing through which allows the lower blocks to swing freely to each side of the boom. In this way the derrick reaches a vertical position. A bullrope easily pulls the derrick to the other side until the weight of the cargo tips the derrick over. The span tackles now have the weight on the other side. The union table is fixed again and the derrick can start its work on the other side. There are also Stülcken with single-pendulum blocks. At this type the cargo hook is detached and the lower and upper cargo block are hauled into the centre of the Stülcken. To tip the derrick over the gravity is here used again. A derrick helps pump oil in most offshore rigs.

Stand (Drill Pipe)

A Stand (of drill pipe) is two or three joints of drill pipe connected together and *stood* in the derrick vertically, usually while tripping pipe. A stand of collars is similar, only made up of collars and a collar head. The collar head is screwed into the collar to allow it to be picked up by the elevators.

Stands are emplaced inside of the "board" of the drilling rig. They are usually kept between "fingers". Most boards will allow stands to go ten stands deep and as much as fifty stands wide on land based rigs. The stands are further held in place using ropes in the board which are tied in a shoe knot by the derrickman.

Stands are emplaced on the floor of the drilling rig by the chain hand. When stands are being put onto the floor the chainhand is said to be "racking stands". After the bottom of the stand is placed on the floor, the derrickman will unlatch the elevators and pull the stand in either with a rope or with just his arms. When stands are being put back into the hole, the derickman will slam the stand into the elevators to force them to latch. The chainhand will brace against the stand to control it when the driller picks it up. This is referred to as "tailing the pipe" as the chain hand will hold the pipe and allow it to semi-drag them back to the hole. The chain hand then passes it off to the tong hand, who then "stabs" the stand into the pipe already in the hole. Rigs are generally sized by how many stands they can hold in their derrick. Most land based rigs are referred to as "triples" because they hold three joints per stand in their derrick. "Singles" generally do not hold any pipe in the derrick and instead require pipe to be laid down during a pipe trip.

Pipe Rack

A Pipe rack is a set of steel framed structures (typically triangularly shaped) that drill pipe is stacked on for storage or when being readied for use on a drilling rig. Pipe rack also refers to the *wooden pad* on the drill floor where the stands of drill pipe are vertically stacked when tripping pipe (pulling pipe out of or running pipe into the wellbore).

Swivel (Drill Rig)

A Swivel is a mechanical device used on a drilling rig that hangs directly under the travelling block and directly above the kelly drive, that provides the ability for the kelly (and subsequently the drill string) to rotate while allowing the travelling block to remain in a stationary rotational position (yet allow vertical movement up and down the derrick) while simultaneously allowing the introduction of drilling fluid into the *drill string.*

Kelly Drive

A kelly drive refers to a type of well drilling device on an oil drilling rig that employs a section of pipe with a polygonal (three-, four-, six-, or eight-sided) or splined outer surface, which passes through the matching polygonal or splined kelly (mating) bushing and rotary table. This bushing is rotated via the rotary table and thus the pipe and the attached drill string turn while the polygonal pipe is free to slide vertically in the bushing as the bit digs the well deeper. When drilling, the drill bit is attached at the end of the drill string and thus the kelly drive provides the means to turn the bit (assuming that a downhole motor is not being used).

The kelly is the polygonal tubing and the *kelly bushing* is the mechanical device that turns the kelly when rotated by the rotary table. Together they are referred to as a *kelly drive.* The upper end of the kelly is screwed into the swivel, using a left-hand thread to preclude loosening from the right-hand torque applied below. The kelly typically is about 10 ft (3 m) longer than the drill pipe segments, thus leaving a portion of newly drilled hole open below the bit after a new length of pipe has been added ("making a connection"), and the drill string has been lowered until the kelly bushing engages again in the rotary table. The kelly hose is the flexible, high-pressure hose connected from the standpipe to a gooseneck pipe on a swivel above the kelly and allows the free vertical movement of the kelly while facilitating the flow of the drilling fluid down the drill string.

Drill Floor

The Drill Floor is the heart of any drilling rig and is also known as the *pad*. This is the area where the drill string begins its trip into the earth. It is traditionally where joints of pipe are assembled, as well as the BHA (bottom hole assembly), drilling bit, and various other tools. This is the primary work location for roughnecks and the driller. The drill floor is located directly under the derrick.

Bell Nipple

A Bell nipple is a section of large diametre pipe fitted to the top of the blowout preventers that the flow line attaches to via a side outlet, to allow the drilling fluid to flow back over the shale shakers to the mud tanks.

Blowout Preventer

A blowout preventer is a large, specialised valve used to seal, control and monitor oil and gas wells. Blowout preventers were developed to cope with extreme erratic pressures and uncontrolled flow (formation kick) emanating from a well reservoir during drilling. Kicks can lead to a potentially catastrophic event known as a blowout. In addition to controlling the downhole (occurring in the drilled hole) pressure and the flow of oil and gas, blowout preventers are intended to prevent tubing (e.g. drill pipe and well casing), tools and drilling fluid from being blown out of the wellbore (also known as bore hole, the hole leading to the reservoir) when a blowout threatens. Blowout preventers are critical to the safety of crew, rig (the equipment system used to drill a wellbore) and environment, and to the monitoring and maintenance of well integrity; thus blowout preventers are intended to be fail-safe devices.

The term BOP (an initialism rather than a spoken acronym, i.e., pronounced B-O-P, not "bop") is used in oilfield vernacular to refer to blowout preventers.

The abbreviated term preventer, usually prefaced by a type (e.g. ram preventer), is used to refer to a single blowout preventer unit. A blowout preventer may also simply be referred to by its type (e.g. ram). The terms blowout preventer, blowout preventer stack and blowout preventer system are commonly used interchangeably and in a general manner to describe an assembly of several stacked blowout preventers of varying type and function, as well as auxiliary components. A typical subsea deepwater blowout preventer system includes components such as electrical and hydraulic lines, control pods, hydraulic accumulators, test valve, kill and choke lines and valves, riser joint, hydraulic connectors, and a support frame.

Two categories of blowout preventer are most prevalent: *ram* and *annular*. BOP stacks frequently utilise both types, typically with at least one annular BOP stacked above several ram BOPs.

Blowout preventers are used at land and offshore rigs, and subsea. Land and subsea BOPs are secured to the top of the wellbore, known as the wellhead. BOPs on offshore rigs are mounted below the rig deck. Subsea BOPs are connected to the offshore rig above by a drilling riser that provides a continuous pathway for the drill string and fluids emanating from the wellbore. In effect, a riser extends the wellbore to the rig.

Use

The invention of blowout preventers was instrumental in reducing the incidence of oil gushers, blowouts, indicating that substantial improvement is needed.

Blowout preventers come in a variety of styles, sizes and pressure ratings. Several individual units serving various functions are combined to compose a blowout preventer stack. Multiple blowout preventers of the same type are frequently provided for redundancy, an important factor in the effectiveness of fail-safe devices.

The primary functions of a blowout preventer system are to:

- Confine well fluid to the wellbore;
- Provide means to add fluid to the wellbore;
- Allow controlled volumes of fluid to be withdrawn from the wellbore.

Additionally, and in performing those primary functions, blowout preventer systems are used to:

- Regulate and monitor wellbore pressure;
- Centre and hang off the drill string in the wellbore;
- Shut in the well (e.g. seal the void, annulus, between drillpipe and casing);

- "Kill" the well (prevent the flow of formation fluid, influx, from the reservoir into the wellbore) ;
- Seal the wellhead (close off the wellbore);
- Sever the casing or drill pipe (in case of emergencies).

In drilling a typical high-pressure well, drill strings are routed through a blowout preventer stack toward the reservoir of oil and gas. As the well is drilled, drilling fluid, "mud", is fed through the drill string down to the drill bit, "blade", and returns up the wellbore in the ring-shaped void, annulus, between the outside of the drill pipe and the casing (piping that lines the wellbore). The column of drilling mud exerts downward hydrostatic pressure to counter opposing pressure from the formation being drilled, allowing drilling to proceed.

When a kick (influx of formation fluid) occurs, rig operators or automatic systems close the blowout preventer units, sealing the annulus to stop the flow of fluids out of the wellbore. Denser mud is then circulated into the wellbore down the drill string, up the annulus and out through the choke line at the base of the BOP stack through chokes (flow restrictors) until downhole pressure is overcome. Once "kill weight" mud extends from the bottom of the well to the top, the well has been "killed". If the integrity of the well is intact drilling may be resumed. Alternatively, if circulation is not feasible it may be possible to kill the well by "bullheading", forcibly pumping, in the heavier mud from the top through the kill line connection at the base of the stack. This is less desirable because of the higher surface pressures likely needed and the fact that much of the mud originally in the annulus must be forced into receptive formations in the open hole section beneath the deepest casing shoe.

If the blowout preventers and mud do not restrict the upward pressures of a kick, a blowout results, potentially shooting tubing, oil and gas up the wellbore, damaging the rig, and leaving well integrity in question. Since BOPs are important for the safety of the crew and natural environment, as well as the drilling rig and the wellbore itself, authorities recommend, and regulations require, that BOPs be regularly inspected, tested and refurbished. Tests vary from daily test of functions on critical wells to monthly or less frequent testing on wells with low likelihood of control problems.

Exploitable reservoirs of oil and gas are increasingly rare and remote, leading to increased subsea deepwater well exploration and requiring BOPs to remain submerged for as long as a year in extreme conditions. As a result, BOP assemblies have grown larger and heavier (e.g. a single ram-type BOP unit can weigh in excess of 30,000 pounds), while the space allotted for BOP stacks on existing offshore rigs has not grown commensurately. Thus a key focus in the technological development of BOPs over the last two decades has been limiting their footprint and weight while simultaneously increasing safe operating capacity.

Types

BOPs come in two basic types, *ram* and *annular*. Both are often used together in drilling rig BOP stacks, typically with at least one annular BOP capping a stack of several ram BOPs.

Ram Blowout Preventer

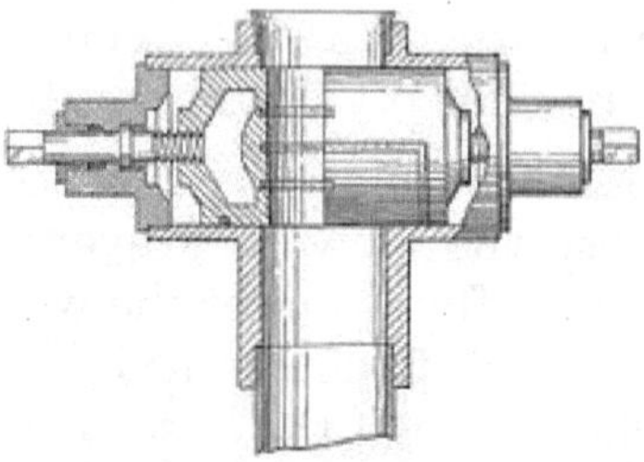

Figure: *A Patent Drawing of the Original Ram-type Blowout Preventer, by Cameron Iron Works (1922).*

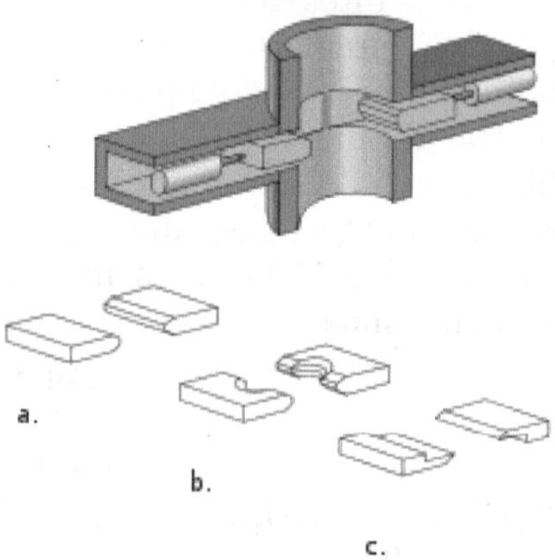

Figure: *Blowout Preventer diagram showing different types of rams. (a) blind ram (b) pipe ram and (c) shear ram.*

The ram BOP was invented by James Smither Abercrombie and Harry S. Cameron in 1922, and was brought to market in 1924 by Cameron Iron Works.

A ram-type BOP is similar in operation to a gate valve, but uses a pair of opposing steel plungers, rams. The rams extend toward the centre of the wellbore to restrict flow or retract open in order to permit flow. The inner and top faces of the rams are fitted with packers (elastomeric seals) that press against each other, against the wellbore, and around tubing running through the wellbore. Outlets at the sides of the BOP housing (body) are used for connection to choke and kill lines or valves.

Rams, or ram blocks, are of four common types: *pipe*, *blind*, *shear*, and *blind shear*.

Pipe rams close around a drill pipe, restricting flow in the annulus (ring-shaped space between concentric objects) between the outside of the drill pipe and the wellbore, but do not obstruct flow within the drill pipe. Variable-bore pipe rams can accommodate tubing in a wider range of outside diametres than standard pipe rams, but typically with some loss of pressure capacity and longevity. Blind rams (also known as sealing rams), which have no openings for tubing, can close off the well when the well does not contain a drill string or other tubing, and seal it.

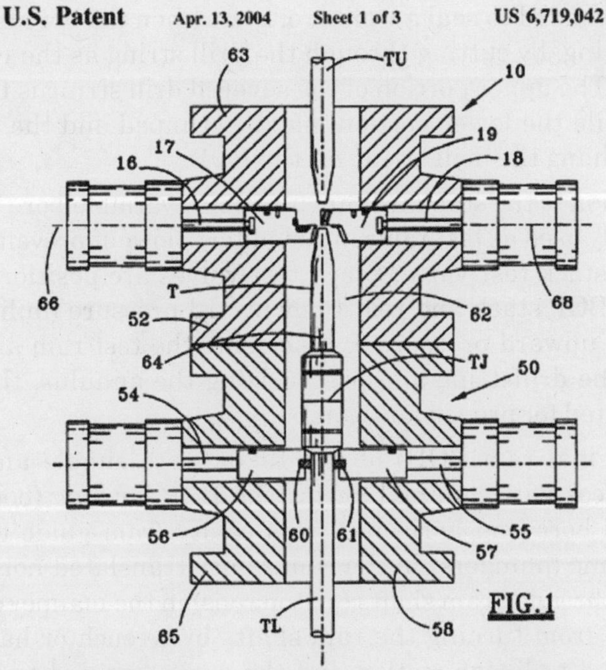

10 Shear Ram Preventer
16 Shear Ram Shaft
17 Lower Blade Shear Ram
18 Shear Ram Shaft
19 Upper Blade Shear Ram
50 Pipe Ram Preventer
52 Housing (Body)
54 Pipe Ram Shaft
55 Pipe Ram Shaft
56 Pipe Ram (Ram Block)
57 Pipe Ram (Ram Block)
58 Housing (Body)
60 Packer (Ram Seal)
61 Packer (Ram Seal)
62 Wellbore
63 Upper Flange
64 Wellbore Axis
65 Lower Flange
66 Ram Axis
68 Ram Axis
T Tubular
TJ Tool Joint
TL Lower Tubular
TU Upper Tubular

Figure: *Patent Drawing of a Varco Shaffer Ram BOP Stack. A shear ram BOP has cut the drillstring and a pipe ram has hung it off.*

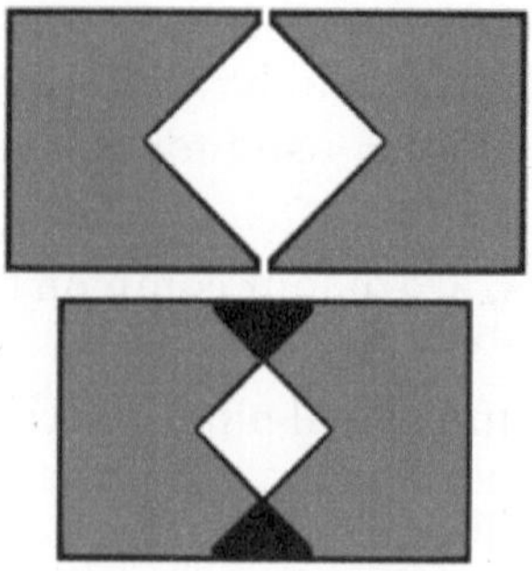

Figure: *Schematic view of closing shear blades*

Shear rams cut through the drill string or casing with hardened steel shears.

Blind shear rams (also known as shear seal rams, or sealing shear rams) are intended to seal a wellbore, even when the bore is occupied by a drill string, by cutting through the drill string as the rams close off the well. The upper portion of the severed drill string is freed from the ram, while the lower portion may be crimped and the "fish tail" captured to hang the drill string off the BOP.

In addition to the standard ram functions, variable-bore pipe rams are frequently used as test rams in a modified blowout preventer device known as a stack test valve. Stack test valves are positioned at the bottom of a BOP stack and resist downward pressure (unlike BOPs, which resist upward pressures). By closing the test ram and a BOP ram about the drillstring and pressurizing the annulus, the BOP is pressure-tested for proper function.

The original ram BOPs of the 1920s were simple and rugged manual devices with minimal parts. The BOP housing (body) had a vertical well bore and horizontal ram cavity (ram guide chamber). Opposing rams (plungers) in the ram cavity translated horizontally, actuated by threaded ram shafts (piston rods) in the manner of a screw jack. Torque from turning the ram shafts by wrench or hand wheel was converted to linear motion and the rams, coupled to the inner ends of the ram shafts, opened and closed the well bore. Such screw jack type operation provided enough mechanical advantage for rams to overcome downhole pressures and seal the wellbore annulus.

Hydraulic rams BOPs were in use by the 1940s. Hydraulically actuated blowout preventers had many potential advantages. The pressure could be equalized in the opposing hydraulic cylinders causing the rams to operate in unison. Relatively rapid actuation and remote control were facilitated, and hydraulic rams were well-suited to high pressure wells.

Because BOPs are fail-safe devices, efforts to minimise the complexity of the devices are still employed to ensure ram BOP reliability and longevity. As a result, despite the ever-increasing demands placed on them, state of the art ram BOPs are conceptually the same as the first effective models, and resemble those units in many ways.

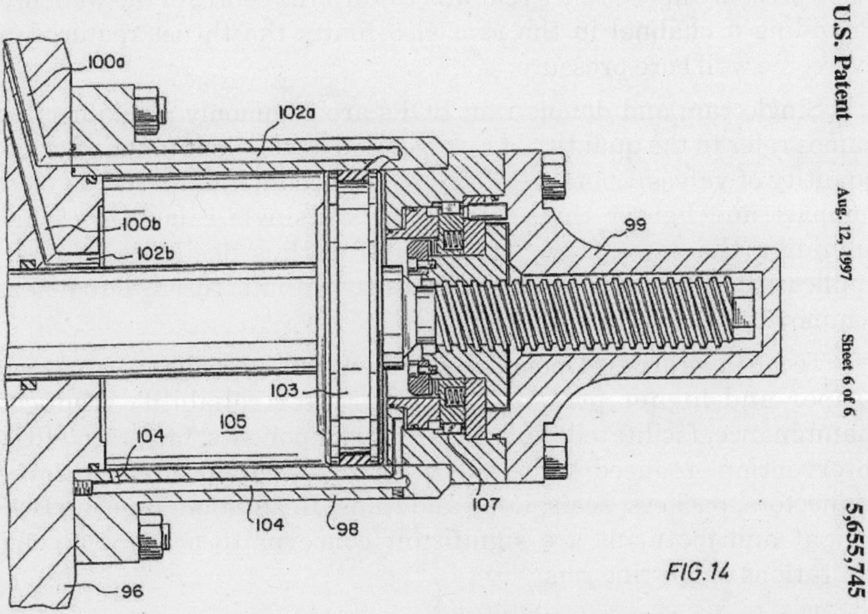

Figure: *Hydril Company's Compact BOP Ram Actuator Assembly Patent Drawing*

Ram BOPs for use in deepwater applications universally employ hydraulic actuation. Threaded shafts are often still incorporated into hydraulic ram BOPs as lock rods that hold the ram in position after hydraulic actuation. By using a mechanical ram locking mechanism, constant hydraulic pressure need not be maintained. Lock rods may be coupled to ram shafts or not, depending on manufacturer. Other types of ram locks, such as wedge locks, are also used.

Typical ram actuator assemblies (operator systems) are secured to the BOP housing by removable bonnets. Unbolting the bonnets from the housing allows BOP maintenance and facilitates the substitution of rams. In that way, for example, a pipe ram BOP can be converted to a blind shear ram BOP.

Shear-type ram BOPs require the greatest closing force in order to cut through tubing occupying the wellbore. Boosters (auxiliary hydraulic actuators) are frequently mounted to the outer ends of a BOP's hydraulic actuators to provide additional shearing force for shear rams.

Ram BOPs are typically designed so that well pressure will help maintain the rams in their closed, sealing position. That is achieved by allowing fluid to pass to pass through a channel in the ram and exert pressure at the ram's rear and toward the centre of the wellbore. Providing a channel in the ram also limits the thrust required to overcome well bore pressure.

Single ram and double ram BOPs are commonly available. The names refer to the quantity of ram cavities (equivalent to the effective quantity of valves) contained in the unit. A double ram BOP is more compact and lighter than a stack of two single ram BOPs while providing the same functionality, and is thus desirable in many applications. Triple ram BOPs are also manufactured, but not as common.

Technological development of ram BOPs has been directed towards deeper and higher pressure wells, greater reliability, reduced maintenance, facilitated replacement of components, facilitated ROV intervention, reduced hydraulic fluid consumption, and improved connectors, packers, seals, locks and rams. In addition, limiting BOP weight and footprint are significant concerns to account for the limitations of existing rigs.

The highest-capacity large-bore ram blowout preventer on the market, as of July 2010, Cameron's EVO 20K BOP, has a hold-pressure rating of 20,000 psi, ram force in excess of 1,000,000 pounds, and a well bore diametre of 18.75 inches.

Annular Blowout Preventer

The annular blowout preventer was invented by Granville Sloan Knox in 1946; a U.S. patent for it was awarded in 1952. Often around the rig it is called the "Hydril", after the name of one of the manufacturers of such devices.

An annular-type blowout preventer can close around the drill string, casing or a non-cylindrical object, such as the kelly. Drill pipe including the larger-diametre tool joints (threaded connectors) can be "stripped" (i.e., moved vertically while pressure is contained below) through an annular preventer by careful control of the hydraulic closing pressure. Annular blowout preventers are also effective at maintaining a seal around the drillpipe even as it rotates during drilling. Regulations typically require that an annular preventer be able to completely close a wellbore, but annular preventers are generally not as effective as ram preventers in maintaining a seal on an open hole.

Annular BOPs are typically located at the top of a BOP stack, with one or two annular preventers positioned above a series of several ram preventers.

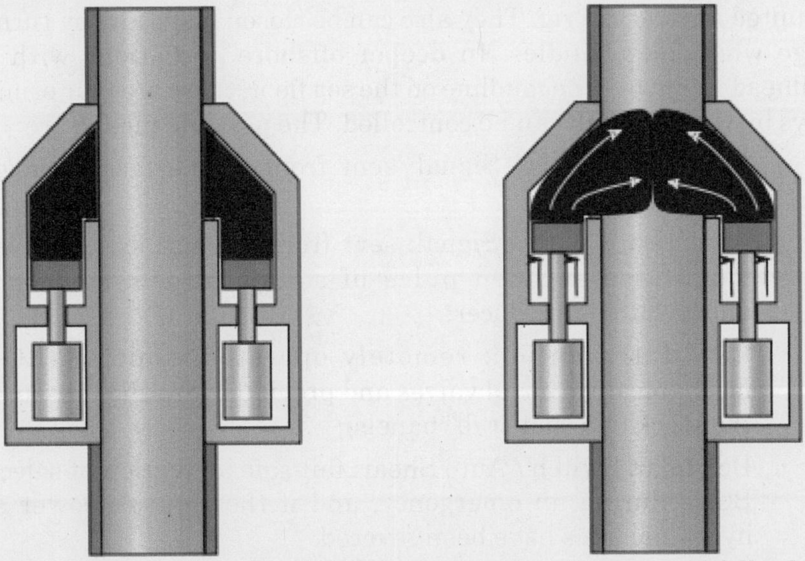

Figure: *Diagram of an Annular blowout preventer in open and fully closed configurations. The flexible annulus (donut) in blue is forced into the drillpipe cavity by the hydraulic pistons.*

An annular blowout preventer uses the principle of a wedge to shut in the wellbore. It has a donut-like rubber seal, known as an elastomeric packing unit, reinforced with steel ribs. The packing unit is situated in the BOP housing between the head and hydraulic piston. When the piston is actuated, its upward thrust forces the packing unit to constrict, like a sphincter, sealing the annulus or openhole. Annular preventers have only two moving parts, piston and packing unit, making them simple and easy to maintain relative to ram preventers.

The original type of annular blowout preventer uses a “wedge-faced” (conical-faced) piston. As the piston rises, vertical movement of the packing unit is restricted by the head and the sloped face of the piston squeezes the packing unit inward, toward the centre of the wellbore. In 1972, Ado N. Vujasinovic was awarded a patent for a variation on the annular preventer known as a spherical blowout preventer, so-named because of its spherical-faced head. As the piston rises the packing unit is thust upward against the curved head, which constricts the packing unit inward. Both types of annular preventer are in common use.

Control Methods

When rigs are drilled on land or in very shallow water where the wellhead is above the water line, BOPs are activated by hydraulic pressure from a remote accumulator. Several control stations will be mounted around the rig. They also can be closed manually by turning large wheel-like handles. In deeper offshore operations with the wellhead just above the mudline on the sea floor, there are four primary ways by which a BOP can be controlled. The possible means are:

- Electrical Control Signal: sent from the surface through a control cable;
- Acoustical Control Signal: sent from the surface based on a modulated/encoded pulse of sound transmitted by an underwater transducer;
- ROV Intervention: remotely operated vehicles (ROVs) mechanically control valves and provide hydraulic pressure to the stack (via "hot stab" panels);
- Deadman Switch / Auto Shear: fail-safe activation of selected BOPs during an emergency, and if the control, power and hydraulic lines have been severed.

Two control pods are provided on the BOP for redundancy. Electrical signal control of the pods is primary. Acoustical, ROV intervention and dead-man controls are secondary.

An emergency disconnect system, or EDS, disconnects the rig from the well in case of an emergency. The EDS is also intended to automatically trigger the deadman switch, which closes the BOP, kill and choke valves. The EDS may be a subsystem of the BOP stack's control pods or separate.

Pumps on the rig normally deliver pressure to the blowout preventer stack through hydraulic lines. Hydraulic accumulators are on the BOP stack enable closure of blowout preventers even if the BOP stack is disconnected from the rig. It is also possible to trigger the closing of BOPs automatically based on too high pressure or excessive flow.

Individual wells along the U.S. coastline may also be required to have BOPs with backup acoustic control. General requirements of other nations, including Brazil, were drawn to require this method. BOPs featuring this method may cost as much as US$500,000 more than those that omit the feature.

Deepwater Horizon Blowout

During the *Deepwater Horizon* drilling rig explosion incident on April 20, 2010, the blowout preventer should have been activated automatically, cutting the drillstring and sealing the well to preclude a blowout and subsequent oil spill in the Gulf of Mexico, but it failed to fully engage. Underwater robots (ROVs) later were used to manually trigger the blind shear ram preventer, to no avail.

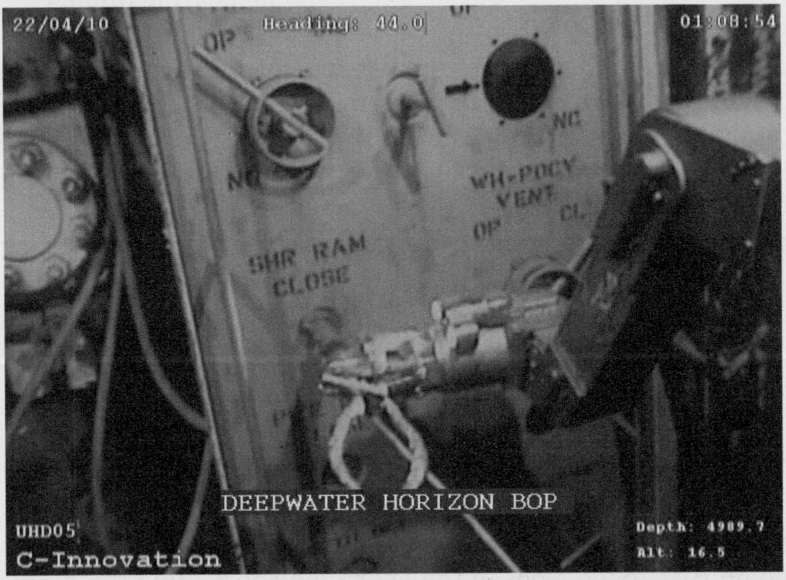

Figure: *A robotic arm of a Remotely Operated Vehicle (ROV) attempts to activate the* "Deepwater Horizon" *Blowout Preventer (BOP), Thursday, April 22, 2010.*

As of May 2010 it is unknown why the blowout preventer failed. Chief surveyor John David Forsyth of the American Bureau of Shipping testified in hearings before the Joint Investigation of the Minerals Management Service and the U.S. Coast Guard investigating the causes of the explosion that his agency last inspected the rig's blowout preventer in 2005. BP representatives suggested that the preventer could have suffered a hydraulic leak. Gamma-ray imaging of the preventer conducted on May 12 and May 13, 2010 showed that the preventer's internal valves were partially closed and were restricting the flow of oil. Whether the valves closed automatically during the explosion or were shut manually by remotely operated vehicle work is unknown. A statement released by Congressman Bart Stupak revealed that, among other issues, the emergency disconnect system (EDS) did not function as intended and may have malfunctioned due to the explosion on the Deepwater Horizon.

The permit for the Macondo Prospect by the Minerals Management Service in 2009 did not require redundant acoustic control means. Inasmuch as the BOPs could not be closed successfully by underwater manipulation (ROV Intervention), pending results of a complete investigation, it is uncertain whether this omission was a factor in the blowout.

Documents discussed during congressional hearings June 17, 2010, suggested that a battery in the device's control pod was flat and that the rig's owner, Transocean, may have "modified" Cameron's equipment for the Macondo site (including incorrectly routing hydraulic pressure to a stack test valve instead of a pipe ram BOP) which increased the risk of BOP failure, in spite of warnings from their contractor to that effect. Another hypothesis is that a junction in the drilling pipe may have been positioned in the BOP stack in such way that its shear rams had an insurmountable thickness of material to cut through.

It was later discovered that a second piece of tubing got into the BOP stack at some point during the Macondo incident, potentially explaining the failure of the BOP shearing mechanism. As of July 2010 it is unknown whether the tubing might be casing that shot up through the well or perhaps broken drill pipe that dropped into the well.

On July 10, 2010 BP began operations to install a sealing cap, also known as a capping stack, atop the failed blowout preventer stack. Based on BP's video feeds of the operation the sealing cap assembly, called Top Hat 10, includes a stack of three blind shear ram BOPs manufactured by Hydril (a GE Oil & Gas company), one of Cameron's chief competitors. By July 15 the 3 ram capping stack had sealed the Macondo well, if only temporarily, for the first time in 87 days.

The U.S. government wants the failed blowout preventer to be replaced in case of any pressure that occurs when the relief well intersects with the well. On September 3 at 1:20 p.m. CDT the 300 ton failed blowout preventer was removed from the well and began being slowly lifted to the surface. Later that day a replacement blowout preventer was placed on the well. On September 4 at 6:54 p.m. CDT the failed blowout preventer reached the surface of the water and at 9:16 p.m. CDT it was placed in a special container on board the vessel Helix Q4000. The failed blowout preventer was taken to a NASA facility in Louisiana for examination by Det Norske Veritas (DNV).

On 20 March 2011, DNV presented their report to the US Department of Energy. Their primary conclusion was that the rams failed to shear through the oil pipe and seal it because it has buckled out of the line of action of the rams. They did not suggest any failure of actuation as would be caused by faulty batteries.

Drill String

A drill string on a drilling rig is a column, or string, of drill pipe that transmits drilling fluid (via the mud pumps) and torque (via the kelly drive or top drive) to the drill bit. The term is loosely applied as the assembled collection of the drill pipe, drill collars, tools and drill bit. The drill string is hollow so that drilling fluid can be pumped down through it and circulated back up the annulus (the void between the drill string and the casing/open hole).

Drill String Components

The drill string is typically made up of three sections:

- Bottom hole assembly (BHA)
- Transition pipe, which is often heavyweight drill pipe (HWDP)
- Drill pipe

Bottom Hole Assembly (BHA)

The BHA is made up of: a drill bit, which is used to break up the rock formations; drill collars, which are heavy, thick-walled tubes used to apply weight to the drill bit; and drilling stabilisers, which keep the assembly centred in the hole. The BHA may also contain other components such as a downhole motor and Rotary steerable system, measurement while drilling (MWD), and logging while drilling (LWD) tools. The components are joined together using rugged threaded connections. Short "subs" are used to connect items with dissimilar threads.

Transition Pipe

Heavyweight drill pipe (HWDP) may be used to make the transition between the drill collars and drill pipe. The function of the HWDP is to provide a flexible transition between the drill collars and the drill pipe. This helps to reduce the number of fatigue failures seen directly above the BHA. A secondary use of HWDP is to add additional weight to the drill bit.

Drill Pipe

Drill pipe makes up the majority of the drill string back up to the surface. Each segment has a larger-diametre portion on each end containing a male ("pin") or female ("box") thread.

Running a Drill String

Most components in a drill string are manufactured in 31 foot lengths (range 2) although they can also be manufactured in 46 foot lengths (range 3). Each 31 foot component is referred to as a joint. Typically 2, 3 or 4 joints are joined together to make a stand. Modern onshore rigs are capable of handling ~90 ft stands (often referred to as a triple).

Pulling the drill string out of or running the drill string into the hole is referred to as tripping. Drill pipe, HWDP and collars are typically stood back in stands in the derrick if they are to be run back into the hole again after, say, changing the bit. The disconnect point ("break") is varied each subsequent round trip so that after three trips every connection has been broken apart and later made up again with fresh pipe dope applied.

Stuck Drill String

A stuck drill string can be caused by many situations.

- Packing-off due to cuttings settling back into the wellbore when circulation is stopped.
- Differentially when the formation pressure is too low, the wellbore pressure is too high or both, essentially pushing the pipe onto the wall of the wellbore.
- Keyhole sticking occurs mechanically as a result of pulling up into doglegs when tripping.
- Adhesion due to not moving it for a significant amount of time.

Once the tubular member is stuck, there are many techniques used to extract the pipe. The tools and expertise are normally supplied by an oilfield service company. Two popular tools and techniques are the oilfield jar and the surface resonant vibrator. Below is a history of these tools along with how they operate.

History of Jars

The mechanical success of cable tool drilling has greatly depended on a device called jars, invented by a spring pole driller, William Morris, in the salt well days of the 1830s. Little is known about Morris except for his invention and that he listed Kanawha County (now in West Virginia) as his address. Morris received US 2243 for this unique tool in 1841 for artesian well drilling. Later, using jars, the cable tool system was able to efficiently meet the demands of drilling wells for oil.

The jars were improved over time, especially at the hands of the oil drillers, and reached the most useful and workable design by the

1870s, due to another US 78958 received in 1868 by Edward Guillod of Titusville, Pennsylvania, which addressed the use of steel on the jars' surfaces that were subject to the greatest wear. Many years later, in the 1930s, very strong steel alloy jars were made.

A set of jars consisted of two interlocking links which could telescope. In 1880 they had a play of about 13 inches such that the upper link could be lifted 13 inches before the lower link was engaged. This engagement occurred when the cross-heads came together. Today, there are two primary types, hydraulic and mechanical jars. While their respective designs are quite different, their operation is similar. Energy is stored in the drillstring and suddenly released by the jar when it fires. Jars can be designed to strike up, down, or both. In the case of jarring up above a stuck bottomhole assembly, the driller slowly pulls up on the drillstring but the BHA does not move. Since the top of the drillstring is moving up, this means that the drillstring itself is stretching and storing energy. When the jars reach their firing point, they suddenly allow one section of the jar to move axially relative to a second, being pulled up rapidly in much the same way that one end of a stretched spring moves when released. After a few inches of movement, this moving section slams into a steel shoulder, imparting an impact load.

In addition to the mechanical and hydraulic versions, jars are classified as drilling jars or fishing jars. The operation of the two types is similar, and both deliver approximately the same impact blow, but the drilling jar is built such that it can better withstand the rotary and vibrational loading associated with drilling. Jars are designed to be reset by simple string manipulation and are capable of repeated operation or firing before being recovered from the well. Jarring effectiveness is determined by how rapidly you can impact weight into the jars. When jarring without a compounder or accelerator you rely only on pipe stretch to lift the drill collars upwards after the jar releases to create the upwards impact in the jar. This accelerated upward movement will often be reduced by the friction of the working string along the sides of the well bore, reducing the speed of upwards movement of the drill collars which impact into the jar. At shallow depths jar impact is not achieved because of lack of pipe stretch in the working string.

When pipe stretch alone cannot provide enough energy to free a fish, compounders or accelerators are used. Compounders or accelerators are energized when you over pull on the working string and compress a compressible fluid through a few feet of stroke distance and at the same time activate the fishing jar. When the fishing jar releases the stored energy in the compounder/acclerator lifts the drill collars upwards at a high rate of speed creating a high impact in the jar.

System Dynamics of Jars

Jars rely on the principle of stretching a pipe to build elastic potential energy such that when the jar trips it relies on the masses of the drill pipe and collars to gain velocity and subsequently strike the anvil section of jar. This impact results in a force, or blow, which is converted into energy.

History of Surface Resonant Vibrators

The concept of using vibration to free stuck objects from a wellbore originated in the 1940s, and probably stemmed from the 1930s use of vibration to drive piling in the Soviet Union. The early use of vibration for driving and extracting piles was confined to low frequency operation; that is, frequencies less than the fundamental resonant frequency of the system and consequently, although effective, the process was only an improvement on conventional hammer equipment. Early patents and teaching attempted to explain the process and mechanism involved, but lacked a certain degree of sophistication. In 1961, A. G. Bodine obtained US 2972380 that was to become the "mother patent" for oil field tubular extraction using sonic techniques.

The first published work on this technique was outlined in a 1987 Society of Petroleum Engineers (SPE) paper presented at the International Association of Drilling Contractors in Dallas, Texas detailing the nature of the work and the operational results that were achieved. The cited work involving liner, tubing, and drill pipe extraction and was very successful. Reference Two presented at the Society of Petroleum Engineers Annual Technical Conference and Exhibition in Aneheim, Ca, November, 2007 explains the resonant vibration theory in more detail as well as its use in extracting long lengths of mud stuck tubulars.

System Dynamics of Surface Resonant Vibrators

Surface Resonant Vibrators rely on the principle of counter rotating eccentric weights to impart a sinusoidal harmonic motion from the surface into the work string at the surface. Reference Three (above) provides a full explanation of this technology. The frequency of rotation, and hence vibration of the pipe string, is tuned to the resonant frequency of the system. The system is defined as the surface resonant vibrator, pipe string, fish and retaining media. The resultant forces imparted to the fish is based on the following logic:

- The delivery forces from the surface are a result of the static overpull force from the rig, plus the dynamic force component of the rotating eccentric weights
- Depending on the static overpull force component, the resultant force at the fish can be either tension or compression due to the sinusoidal force wave component from the oscillator
- Initially during startup of a vibrator, some force is necessary to lift and lower the entire load mass of the system. When the vibrator tunes to the resonant frequency of the system, the reactive load impedance cancels out to zero by virtue of the inductance reactance (mass of the system) equalling the compliance or stiffness reactance (elasticity of the tubular). The remaining impedance of the system, known as the resistive load impedance, is what is retaining the stuck pipe.
- During resonant vibration, a longitudinal sine wave travels down the pipe to the fish with an attendant pipe mass that is equal to a quarter wavelength of the resonant vibrating frequency.
- A phenomenon known as fluidization of soil grains takes place during resonant vibration whereby the granular material constraining the stuck pipe is transformed into a fluidic state that offers little resistance to movement of bodies through the media. In effect, it takes on the characteristics and properties of a liquid.
- During pipe vibration, Dilation and Contraction of the pipe body, known as Poisson's ratio, takes place such that when the stuck pipe is subjected to axial strain due to stretching, its diametre will contract. Similarly, when the length of pipe is compressed, its diametre will expand. Since a length of pipe undergoing vibration experiences alternate tensile and compressive forces as waves along its longitudinal axis (and therefore longitudinal strains), its diametre will expand and contract in unison with the applied tensile and compressive waves. This means that for alternate moments during a vibration cycle the pipe may actually be physically free of its bond.

Drill Bit (Well)

A Drill bit, is a device attached to the end of the drill string that breaks apart, cuts or crushes the rock formations when drilling a wellbore, such as those drilled to extract water, gas, or oil.

The drill bit is hollow and has jets to allow for the expulsion of the drilling fluid, or "mud", at high velocity and high pressure to help clean the bit and, for softer formations, help to break apart the rock. A tricone bit comprises three conical rollers with teeth made of a hard material, such as tungsten carbide. The teeth break rock by crushing as the rollers move around the bottom of the borehole. A polycrystalline diamond compact (PDC) bit has no moving parts and works by scraping the rock surface with disk-shaped teeth made of a slug of synthetic diamond attached to a tungsten carbide cylinder. The tricone bit is an improvement on the original bit patented in 1909 by Howard R. Hughes, Sr. of Houston, Texas, father of the famed billionaire Howard R. Hughes, Jr..

PDC bits, first came into widespread use in 1976, used for gas and oil exploration the North Sea. They are effective at drilling shale formations, especially when used in combination with oil-base drilling muds.

Casing Head

In oil drilling, a casing head is a simple metal flange welded or screwed on to the top of the conductor pipe (also known as *drive-pipe*) or the casing and forms part of the wellhead system for the well. It is the primary interface for the surface pressure control equipment, such as the blowout preventers (for well drilling) or the Christmas tree (for well production). The casing head, when installed, is typically tested to very strict pressure and leak-off parameters to insure viability under blowout conditions, before any surface equipment is installed.

Flow Line

A flow line, used on a drilling rig, is a large diametre pipe (typically a section of casing) that is connected to the bell nipple (under the drill floor) and extends to the possum belly (on the mud tanks) and acts as a return line, (for the drilling fluid as it comes out of the hole), to the mud tanks.

Flow Line Components

The flow line will typically have equipment attached to it, such as a flow show, possum belly and the sample box

Flow Show

The flow show *shows* whether drilling fluid is flowing down the flow line or not and any changes in that flow. High-pressure jets are typically attached along its length to dislodge any obstructions (such as drill cuttings collecting in one spot) that may occur.

Possum Belly

The possum belly is used to slow the flow of returning drilling fluid before it hits the shale shakers. This enables the shale shaker to clean the cuttings out of the drilling fluid before it is returned to the pits for circulation.

Sample Box

Another common add on is the sample box. This is a heavy duty rubber hose that is inserted at the end of the flow line and at the other end emplaced into the sample box itself. The sample box is used to capture samples of drill cuttings for geological logging. The box is typically equipped with a raising door that allows the water and cuttings to escape after a sample is collected.

Stinger Line

A stinger line is similar to a flow line, but unlike a flow line is not used to maintain circulation. The stinger line is attached to the blowout preventer to allow for the pressure from a blowout to be released. The stinger line usually will run parallel to the flow line.

Chapter 6

Relief Well

A relief well is a well drilled to intersect an oil or gas well that has experienced a blowout. Specialised liquid, such as heavy (dense) drilling mud followed by cement, can then be pumped down the relief well in order to stop the flow from the reservoir in the damaged well. The first use of a relief well was in Texas in the mid 1930s when one was drilled to pump water into an oil well that had cratered and caught on fire.

Riserless Light Well Intervention

Riserless Light Well Intervention (RLWI) enables the petroleum operators to increase the oil and gas recovery rate from subsea oil wells. The operation is performed through smaller dynamic positioned monohull vessels instead of larger mobile installations like traditional semisubmersible drilling rigs or drillships.

The RLWI enable subsea well intervention without having to use a drilling riser package connected to the subsea stack (blowout preventer system).

The technology is based on wireline well maintenance, where the cable is routed via a subsea lubricator system into the subsea well. Traditional activities are wireline operations for well logging, perforation and installing or pulling equipment like plugs and downhole safety valve inserts. The operational envelope can be extended by use of tractors in horizontal wells.

The recovery rate has traditionally been considerably lower on a subsea well than for a surface platform well due more complicated well intervention and maintenance issues. RLWI enables the operators to perform intervention to increase the recovery rate at reduced time and cost. There is a positive safety aspect by employing RLWI services as no hydrocarbons are transported to the vessel, but flushed back into the well through the lubricator system during normal operations. However, RLWI is a complex operation requiring special expertise and control in all phases of the preparation and implementation.

Oilfield Terminology

Oilfield terminology refers to the jargon used by those working in fields within and related to the petroleum industry. It includes words and phrases describing professions, equipment, and procedures specific to the industry. It may also include slang terms used by oilfield workers to describe the same.

Examples:

- *Bell nipple:* An enlarged pipe at the top of a casing string that serves as a funnel to guide drilling tools into the top of the well.
- *Blowout:* A sudden, uncontrolled release of underground pressure from the well.
- *CSG (Coal Seam Gas) or CBM* (Coal Bed Methane) is methane that forms in the gaps between coal molecules. (not to be confused with csg, an abbreviation for the well casing)
- *Hanger:* Refers to several different tools. The casing hanger is the portion of a wellhead assembly which provides support for the casing string when it is lowered into the wellbore. The liner hanger is used to hang casing liners (casing strings that do not reach the surface) from the internal wall of a previous casing string. The instrument hanger is a downhole tool on which downhole gauges or instruments that are to be temporarily left in the wellbore are attached. The tubing hanger attaches to the topmost tubing joint in the wellhead to support the tubing string.
- Chainhand (also Motorman): An experienced labourer capable of maintaining most parts of the rig. The chainhand is in charge of throwing the chain to make up or break down pipe stands during tripping pipe. They are also responsible for maintaining the motors on the drilling rig (hence "motorman").
- *Cold vent:* An area of the ocean floor where hydrogen sulfide, methane and other hydrocarbon-rich fluid seepage occurs, often in the form of a brine pool. Constitutes a biome supporting several endemic species. More commonly a "cold seep".
- *Core sample:* A cylindrical section of a naturally occurring medium consistent enough to hold a layered structure. Most

cores are obtained by drilling into the medium, for example sediment or rock, with a hollow steel tube called a corer.

- *Derrickman:* The second in command after the driller.
- *Drill string:* A column, or string, of pipe that transmits drilling fluid (via the mud pumps) and torque (via the kelly drive or top drive) to the drill bit.
- *Driller:* The crew supervisor on a drilling rig, working under the drilling superintendent or the corporation the rig is contracted to.
- *Elevator:* A hinged mechanism that is closed around the drill pipe or other drill string components to facilitate lowering them into or lifting them out of the wellbore.
- *Floorhand:* Labourer responsible for the overall maintenance of the rig.
- *Frac job:* Slang term for hydraulic fracturing, a process whereby an oil or gas well is stimulated by pumping special fluids into the reservoir, fracturing the formation.
- *Fish:* Any object that has been unintentionally dropped and lost down the wellbore. Fishing is the act of using specialised tools lowered on the drillstring to retrieve a fish.
- *Ginzel:* Slang term for a worker even lower on the food chain than a worm. A ginzel would be someone just starting out that has no oilfield experience whatsoever.
- Horner plot is a semi-log plot produced during pressure buildup analysis.
- *Intelligent well:* An oil or gas well equipped with monitoring equipment and completion components that allow for automatic or remote optimization of production.
- *Joint:* A length of drill pipe.
- *Kick:* An intrusion of pressurized gas into the wellbore that causes drilling fluid to be displaced. It can be the precursor to a blowout.
- *Kill:* The act of stopping a well from flowing.
- *Well logging:* The practice of making a detailed record (a well log) of the geologic formations penetrated by a borehole. Also known as borehole logging. Wireline logging is the practice of measuring formation properties using electrically powered instruments to infer properties and make decisions about drilling and production operations.

- Moon pool, an air-filled chamber open to the water below.
- *Mud:* Slang term for drilling fluid. A Mud Man is the drilling fluids technician responsible for formulating the mud, while a Mud Logger checks mud cuttings from the drill bit for traces of rock or oil and gas that provide a picture of conditions downhole.
- *Pay:* A reservoir or portion of a reservoir containing hydrocarbons that can be economically produced, i.e. it is capable of "paying" an income. Also referred to as "pay sand" or the "pay zone".
- *Pig:* A device inserted into a pipeline for cleaning purposes. The act of using a pig is called pigging.
- *Piston corers:* advanced piston corer, extended core barrel systems
- *Petroleum play:* (or "play") A group of oil prospects that are controlled by the same set of geological circumstances.
- *Pressure Core Sampler:* A tool designed to retrieve a sediment core sample from depth under pressure.

Deepwater Drilling

Deepwater drilling is the process of oil and gas exploration and production in depths of more than 500 feet. By this definition, there are approximately 600 deepwater wells in the Gulf of Mexico. It has been economically infeasible for many years, but with rising oil prices, more companies are investing in this area. Major companies working in this sector include Diamond Offshore, TransOcean, and Baker Hughes.

There are basically two kinds of mobile deepwater drilling rigs: semi-submersible drilling rigs and drillships. Drilling can also be performed from a fixed-position installation such as a floating spar (platform) or tension-leg platform. Recent industry analysis has estimated that the total expenditure in the global deepwater infrastructure market will reach $145bn in 2011.

Depth in a Well

In the oil and gas industry, depth in a well is the measurement, for any point in that well, of the distance between a reference point or elevation, and that point. It is the most common method of reference for locations in the well, and therefore, in oil industry speech, "depth" also refers to the location itself.

By extension, depth can refer to locations below, or distances from, a reference point or elevation, even when there is no well. In that sense, depth is a concept related to elevation, albeit in the opposite direction. Depth in a well is not necessarily measured vertically or along a straight line.

Because wells are not always drilled vertically, there may be two "depths" for every given point in a wellbore: the measured depth (MD) measured along the path of the borehole, and the true vertical depth (TVD), the absolute vertical distance between the datum and the point in the wellbore. In perfectly vertical wells, the TVD equals the MD; otherwise, the TVD is less than the MD measured from the same datum. Common datums used are ground level (GL), drilling rig floor (DF), rotary table (RT), kelly bushing (KB) and mean sea level (MSL).

The Specification of Depth

Although it is an intuitive concept, depth in a well is the source of much confusion because it is frequently not specified correctly. Absolute depth should always be specified with three components:

- a unit (e.g. m for metre),
- a path (e.g. MD for measured depth)
- and the reference or datum they refer to (e.g. RT for rotary table).

Depth in a Well as Used in the Oil and Gas Industry

- Well depth values taken during the drilling operation are referred to as "drillers depth". The "total depth" for the well, core depths and all analysis of core / mud and other materials from the drilling hole are measured in "drillers depth".
- Well depth values from the wireline loggers operation are referred to as "loggers depth". The loggers depth are typically considered more reliable than the drillers depth.
- The difference between loggers and drillers depth are due to different stretch in the drilling string when drilling, and the wire line entered into the bore hole during wireline logging operations. This difference is estimated and referred to as "core shift". A core from a certain drillers depth is lined up with a wireline log (loggers depth) and structures in the core are compared with the log and matched.
- Sign Convention - Depth increases positive in the downward direction. This may seem intuitive but confusion can arise when using certain references while integrating data from different sources. Workers mapping surfaces typically use elevation which, by convention, increases positive in the upward direction. Be mindful when integrating depth and elevation. For example, shallow wells drilled onshore often encounter reservoir at negative depths when referenced to sea level, mappers would

define these same reservoirs at positive elevations when referenced to sea level.

- Depth increases in the "down" direction, so an elevation is a negative depth. There is no need to specify above or below the depth reference.

Example: RT = -10 mMDLAT

- The term "subsea" (SS) by itself should not be used, as it is ambiguous. It could mean: below sea floor or bottom, below mean sea level (MSL), below lowest astronomical tide (LAT), etc.
- The acronym TVDSS is commonly used in the oil industry to represent TVD minus the elevation above mean sea level of the depth reference point of the well. The depth reference point is the kelly bushing in the United States and a few other nations, but is the drill floor in most places.
- Differential (or relative) depths or thicknesses should generally be specified with at least two components: a unit and a path, plus any eventual specifiers to remove any possible ambiguity. No specifier should ever be left "implicit" or "understood". There are cases where a path is not needed and in fact should not be specified, because it is defined by the specifier, e.g. isochore (true stratigraphic thickness, independent of well path or inclination).
- It is important to remember that depths, whether "absolute", "relative", "true", etc, have an intrinsic uncertainty and are never really true.
- The distinction between "loggers' depth" and "drillers' depth" is becoming blurred due to the increasing use of logs acquired while drilling (LWD). At the time of writing, the common practice remains that the petrophysicists or geologists define the "official depths" in a well, and these depths are frequently different from the "drillers' depth", after various corrections, tie-ins, etc, have been applied.
- Petrophysicists and drilling operations tend to express depths with reference to the rotary table or the original drill floor; geologists tend to use a common datum such as the mean sea level; geophysicists use the mean sea level. This can introduce much confusion when a unit is not specified with all 3 components: unit, path, and reference.
- Special consideration must be given to depth measurement in toe-up laterals (J-profile). In these cases the measured depth will continue to increase while true vertical depth will decrease toward the toe of the wellbore.

Depth in Practice: Examples from in the Oil and Gas Industry in Australia

- Unit: the usual unit of depth is the metre (m).
- Path: common expressions of path are measured depth (MD) – elsewhere often known as along hole depth (AHD) – and true vertical depth (TVD). Note that using TV for true vertical depth is not consistent with the use of MD for measured depth, hence the recommended TVD.
- Reference:
  - o the legal datum offshore Australia is Lowest Astronomical Tide (LAT). Note that this requirement in itself can cause difficulties as it is difficult to measure offshore and can vary greatly between locations and even with time. There is, however, an advantage to this convention: tidal corrections should always be of the same sign (negative depth), i.e. the sea level is always higher than or equal to LAT.
  - o A commonly used alternative is Mean Sea Level (MSL).
  - o A datum used in the past was Mean Indian Spring Low Water (MILSW). It is comparable to, but not exactly the same as, LAT.
  - o Common references used in operations include: Rotary Table (RT), Drill Floor (DF), Kelly Bushing (KB), Sea Bottom (SB), Ground Level (GL), Casing Bowl Flange (CBF).

Specification of an absolute depth: in Figure above, point P1 might be at 3207 mMDRT and 2370 mTVDMSL, while point P2 might be at 2530 mMDRT and 2502 mTVDLAT.

Specification of a differential depth or a thickness: the thickness of the reservoir penetrated by the well might be 57 mMD or 42 mTVD, even though the reservoir true stratigraphic thickness in that area (or isopach) might be only 10 m, and its true vertical thickness (isochore), 14 m.

Specification of an absolute depth: in Figure above, point P1 might be at 3207 mMDRT and 2370 mTVDMSL, while point P2 might be at 2530 mMDRT and 2502 mTVDLAT.

Specification of a differential depth or a thickness: in Figure above, the thickness of the reservoir penetrated by the well might be 57 mMD or 42 mTVD, even though the reservoir true stratigraphic thickness in that area (or isopach) might be only 10 m, and its true vertical thickness (isochore), 14 m.

Microbial Enhanced Oil Recovery

Microbial Enhanced Oil Recovery (MEOR) is a biological based technology consisting in manipulating function or structure, or both, of microbial environments existing in oil reservoirs. The ultimate aim of MEOR is to improve the recovery of oil entrapped in porous media while increasing economic profits. MEOR is a tertiary oil extraction technology allowing the partial recovery of the commonly residual two-thirds of oil. thus increasing the life of mature oil reservoirs.

MEOR is a multidisciplinary field incorporating, among others: geology, chemistry, microbiology, fluids mechanics, petroleum engineering, environmental engineering and chemical engineering. The microbial processes proceeding in MEOR can be classified according to the oil production problem in the field:

- *well bore clean up* removes mud and other debris blocking the channels where oil flows through;
- *well stimulation* improves the flow of oil from the drainage area into the well bore; and
- *enhanced water floods* increase microbial activity by injecting selected microbes and sometimes nutrients. From the engineering point of view, MEOR is a system integrated by the reservoir, microbes, nutrients and protocol of well injection.

MEOR Outcomes

So far, the outcomes of MEOR are explained based on two predominant rationales:

Increment in oil production: This is done by modifying the interfacial properties of the system oil-water-minerals, with the aim of facilitating oil movement through porous media. In such a system, microbial activity affects fluidity (viscosity reduction, miscible flooding); displacement efficiency (decrease of interfacial tension, increase of permeability); sweep efficiency (mobility control, selective plugging) and driving force (reservoir pressure).

Upgrading. In this case, microbial activity acts may promote the degradation of heavy oils into lighter ones. Alternatively, it can promote desulphurization due to denitrification as well as the removal of heavy metals.

Relevance

Several decades of research and successful applications support the claims of MEOR as a mature technology. Despite those facts, disagreement still exists. Successful stories are specific for each MEOR field application, and published information regarding supportive economical advantages is however inexistent. Despite this, there is consensus considering MEOR one of the cheapest existing EOR methods. However, obscurity exists on predicting whether or not the deployment of MEOR will be successful.

MEOR is, therefore, one of the future research areas with great priority as identified by the "Oil and Gas in the 21st Century Task Force". This is probably because MEOR is a complementary technology that may help recover the 377 billion barrels of oil that are unrecoverable by conventional technologies.

Bias

Before the advent of environmental molecular microbiology, the word "bacteria" was utilised indistinctively in many fields to refer to uncharacterised microbes, and such systematic error affected several disciplines. Therefore, the word "microbe" or "microorganism" will therefore be preferred hereafter in the text.

History

It was in 1926 when Beckam proposed the utilisation of microorganisms as agents for recovering the remnant oil entrapped in porous media. Since that time numerous investigations have been developed, and are extensively reviewed. In 1947, ZoBell and colleagues set the basis of petroleum microbiology applied to oil recovery, whose contribution would be useful for the first MEOR patent granted to Updegraff and colleagues in 1957 concerning the in situ production of oil recovery agents such as gases, acids, solvents and biosurfactants from microbial degradation of molasses. In 1954, the first field test was carried out in the Lisbon field in Arkansas, USA. During that time, Kuznetsov discovered the microbial gas production from oil. From this year and until the 1970s there was intensive research in USA, USSR, Czechoslovakia, Hungary and Poland. The main type of field experiments developed in those countries consisted in injecting exogenous microbes. In 1958, selective plugging with microbial produced biomass was proposed by Heinningen and colleagues. The oil crisis of 1970 triggered a great interest in active MEOR research in more than 15 countries. From 1970 to 2000, basic MEOR research focused on microbial ecology and characterisation of oil reservoirs. In 1983, Ivanov and colleagues developed the strata microbial activation technology. By 1990, MEOR achieved an interdisciplinary technology status. In 1995, a survey of MEOR projects (322) in the USA showed that 81% of the projects successfully increased oil production, and there was not a single case of reduced oil production. Today, MEOR is gaining attention owing to the high prices of oil and the imminent ending of this resource. As a result, several countries are willing to use MEOR in one third of their oil recovery programs by 2010.

MEOR Advantages

There is a plethora of reviewed claims regarding the advantages of MEOR. However, they should be cautiously regarded due to the lack of published supportive evidence. In addition, assessments of both full live cycle analysis and environmental impact are also unknown.

Advantages can be summarised as follows:

- Injected microbes and nutrients are cheap; easy to handle in the field and independent of oil prices.
- Economically attractive for mature oil fields before abandonment.
- Increases oil production.
- Existing facilities require slight modifications.
- Easy application.
- Less expensive set up.
- Low energy input requirement for microbes to produce MEOR agents.
- More efficient than other EOR methods when applied to carbonate oil reservoirs.
- Microbial activity increases with microbial growth. This is opposite to the case of other EOR additives in time and distance.
- Cellular products are biodegradable and therefore can be considered environmentally friendly.

MEOR Disadvantages

- The oxygen deployed in aerobic MEOR can act as corrosive agent on non-resistant topside equipment and down-hole piping
- Anaerobic MEOR requires large amounts of sugar limiting its applicability in offshore platforms due to logistical problems
- Exogenous microbes require facilities for their cultivation.
- Indigenous microbes need a standardized framework for evaluating microbial activity, e.g. specialised coring and sampling techniques.

- Microbial growth is favoured when: layer permeability is greater than 50 md; reservoir temperature is inferior to 80°C, salinity is below 150 g/L and reservoir depth is less than 2400m.

The Environment of an Oil Reservoir

Oil reservoirs are complex environments containing living (microorganisms) and non living factors (minerals) which interact with each other in a complicated dynamic network of nutrients and energy fluxes. Since the reservoir is heterogeneous, so do the variety of ecosystems containing diverse microbial communities, which in turn are able to affect reservoir behaviour and oil mobilisation.

Microbes are living machines whose metabolites, excretion products and new cells may interact with each other or with the environment, positively or negatively, depending on the global desirable purpose, e.g. the enhancement of oil recovery. All these entities, i.e. enzymes, extracellular polymeric substances (EPS) and the cells themselves, may participate as catalyst or reactants. Such complexity is increased by the interplay with the environment, the later playing a crucial role by affecting cellular function, i.e. genetic expression and protein production. Despite this fundamental knowledge on cell physiology, a solid understanding on function and structure of microbial communities in oil reservoirs, i.e. ecophysiology, remains inexistent.

Environmental Constraints

Several factors concomitantly affect microbial growth and activity. In oil reservoirs, such environmental constraints permit the establishment of criteria to assess and compare the suitability of various microorganisms. Those constraints may not be as harsh as other environments on Earth. For example with, connate brines the salinity is higher than that of sea water but lower than that of salt lakes. In addition, pressures up to 20 MPa and temperatures up to 80 °C, in oil reservoirs, are within the limits for the survival of other microorganisms. Some environmental constraints creating selective pressures on cellular systems that may also affect microbial communities in oil reservoirs are:

Temperature

Enzymes are biological catalysts whose function is affected by a variety of factors including temperature, which, at different ranges, may improve or hamper enzymatic mediated reactions. This will have an effect over the optimal cellular growth or metabolism. Such dependency permits classification of microbes according to the range of temperatures at which they grow. For instance: psychrophiles (<25 °C), mesophiles (25-45 °C), thermophiles (45-60 °C) and hyperthermophiles (60-121 °C). Although such cells optimally grow in these temperature ranges there may not be a direct relationship with the production of specific metabolites.

Pressure

Direct Effects

The effects of pressure on microbial growth under deep ocean conditions were investigated by ZoBell and Johson in 1949. They called those microbes whose growth was enhanced by increasing pressure, barophilic. Other classifications of microorganisms are based on whether microbial growth is inhibited at standard conditions (piezophiles) or above 40 MPa (piezotolerants). From a molecular point of view, the review of Daniel shows that at high pressures the DNA double helix becomes denser, and therefore both gene expression and protein synthesis are affected.

Indirect Effect

Increasing pressure increases gas solubility, and this may affect the redox potential of gases participating as electron acceptors and donors, such as hydrogen or CO_2.

Pore Size/Geometry

One study has concluded that substantial bacterial activity is achieved when there are interconnections of pores having at least 0.2µ diametre. It is expected that pore size and geometry may affect chemotaxis. However, this has not been proven at oil reservoir conditions.

pH

The acidity of alkalinity has an impact over several aspects in living and non living systems. For instance:

Surface Charge

Changes in cellular surface and membrane thickness may be promoted by pH due to its ionization power of cellular membrane embedded proteins. The modified ionic regions may interact with mineral particles and affect the motion of cells through the porous media.

Enzymatic Activity

Embedded cell proteins play a fundamental roll in the transport of chemicals across the cellular membrane. Their function is strongly dependent on their state of ionisation, which is in turn strongly affected by pH. In both cases, this may happen in isolated or complex environmental microbial communities. So far the understanding on the interaction between pH and environmental microbial communities remains unknown, despite the efforts of the last decade. Little is know on the ecophysiology of complex microbial communities and research is still in developmental stage..

Oxidation Potential

The oxidation potential (Eh, measured in volts) is, as in any reaction system, the thermodynamic driving force of anaerobic respiration, which takes place in oxygen depleted environments. Prokaryotes are among the cells that have anaerobic respiration as metabolic strategy for survival. The electron transport takes place along and across the cellular membrane (prokaryotes lack of mitochondria). Electrons are transferred from an electron donor (molecule to be oxidised anaerobically) to an electron acceptor (NO_3, SO_4, MnO_4, etc.). The net Eh between a given electron donor and acceptor; hydrogen ions and other species in place will determine which reaction will first take place. For instance, nitrification is hierarchically more favoured than sulphate reduction. This allows for enhanced oil recovery by disfavouring biologically produced H_2S, which derives from reduced SO_4. In this process, the effects of nitrate reduction on wettability, interfacial tension, viscosity, permeability, biomass and biopolymer production remain unknown.

Electrolyte Composition

Electrolytes concentration and other dissolved species may affect cellular physiology. Dissolving electrolytes reduces thermodynamic activity (aw), vapour pressure and autoprotolysis of water. Besides, electrolytes promote an ionic strength gradient across cellular membrane and therefore provides a powerful driving force allowing the diffusion of water into or out to cells. In natural environments, most bacteria are incapable of living at aw below 0.95. However, some microbes from hypersaline environment such as Pseudomonas species and Halococcus thrive at lower a_w, and are therefore interesting for MEOR research.

Non-Specific Effects

They may occur on pH and Eh. For example, increasing ionic strength increases solubility of nonelectrolytes ('salting out') as in the case of dissolution of carbon dioxide, a pH controller of a variety of natural waters.

Biological Factors

Although it is widely accepted that predation, parasitism, syntrophism and other relationships also occur in the microbial world, little is known in this relationships on MEOR and they have been disregarded in MEOR experiments. In other cases, some microorganisms can thrive in nutrient deficient environments (oligotrophy) such as deep granitic and basaltic aquifers. Other microbes, living in sediments, may utilise available organic compounds (heterotrophy). Organic matter and metabolic products between geological formations can diffuse and support microbial growth in distant environments.

MEOR Mechanism

Understanding MEOR mechanism is still far from being clear. Although a variety of explanations has been given in isolated experiments, it is unclear if they were carried out trying to mimic oil reservoirs conditions.

The mechanism can be explained from the client-operator viewpoint which considers a series of concomitant positive or negative effects that will result in a global benefit:

- *Beneficial effects.* Biodegradation of big molecules reduces viscosity; production of surfactants reduces interfacial tension; production of gas provides additional pressure driving force; microbial metabolites or the microbes themselves may reduce permeability by activation of secondary flow paths.
- *Detrimental effects.* Biologically produced hydrogen sulphide, i.e. souring, causes corrosion of piping and machinery; consumption of hydrocarbons by bacteria reduces the production of desired chemicals.
- *Beneficial or Detrimental.* Permeability reduction can be beneficial in some cases but detrimental in others. Negatively, microbial metabolites or the microbes themselves may reduce permeability by activation of secondary flow paths by depositing: biomass (biological clogging), minerals (chemical clogging) or other suspended particles (physical clogging). Positively, attachment of bacteria and development of slime, i.e. extracellular polymeric substances (EPS), favour the plugging of highly permeable zones (thieves zones) leading to increased sweep efficiency.

Biomass and Biopolymers

In selective plugging, conditioned cells and extracellular polymeric substances plug high permeability zones, resulting in a change of direction of the water flood to oil-rich channels, consequently increasing the sweep efficiency of oil recovery with water flooding. Biopolymer production and the resulting biofilm formation (less 27% cells, 73-98% EPS and void space) are affected by water chemistry, pH, surface charge, microbial physiology, nutrients and fluid flow.

Biosurfactants

Microbial produced surfactants, i.e. biosurfactants reduce the interfacial tension between water and oil, and therefore a lower hydrostatic pressure is required to move the liquid entrapped in the pores to overcome the capillary effect. Secondly, biosurfactants contribute to the formation of micelles providing a physical mechanism to mobilise oil in a moving aqueous phase. Hydrophobic and hydrophilic compounds are in play and have attracted attention in MEOR research, and the main structural types are lipopeptides and glycolipids, being the fatty acid molecule the hydrophobic part.

Gas and Solvents

In this old practice, the production of gas has a positive effect in oil recovery by increasing the differential pressure driving the oil movement. Anaerobically produced methane from oil degradation have a low effect on MEOR due to its high solubility at high pressures. Carbon dioxide is also a good MEOR agent. The miscible CO_2 is condensed into the liquid phase when light hydrocarbons are vaporised into the gas phase. Immiscible CO_2 helps to saturate oil, resulting in swelling and reduction of viscosity of the liquid phase and consequently improving mobilisation by extra driving pressure. Concomitantly, other gases and solvents may dissolve carbonate rock, leading to an increase in rock permeability and porosity.

Field Studies

Worldwide MEOR field applications have been reviewed in detail. Although the exact number field trials is unknown, Lazar et al. suggested an order of hundreds. Successful MEOR field trials have been conducted in the U.S., Russia, China, Australia, Argentina, Bulgaria, former Czechoslovakia, former East Germany, Hungary, India, Malaysia, Peru, Poland, and Romania. Lazar et al. suggested China is leading in the area, and also found that the most successful study was carried out in Alton field, Australia (40% increase of oil production in 12 months).

The majority of the field trials were done in sandstone reservoirs and very few in fractured reservoirs and carbonates. The only known offshore field trials were in Norne (Norway) and Bokor (Malaysia).

As reviewed by Lazar et al., field application followed different approaches such as injection of exogenous microorganisms (microbial flooding); control of paraffin deposition; stimulation of indigenous microbes; injection of ex situ produced biopolymers; starved selected ultramicrobes (selected plugging); selected plugging by sand consolidation due to biomineralization and fracture clogging in carbonate formations; nutrient manipulation of indigenous reservoir microbes to produce ultramicrobes; and adapted mixed enrichment cultures. Reported MEOR results from field trials vary widely. Rigorous controlled experiments are lacking and may not be possible due to the dynamic changes in the reservoir when oil is being recovered. Besides, the economical advantages of these field trials are unknown, and the answer to why the other trials were unsuccessful is unknown. General conclusions can not be drawn because the physical and mineralogical characteristics of the oil reservoirs reported were different. The extrapolation of such conclusions is therefore unviable.

Models

A plethora of attempts to model MEOR has been published. Until now, it is unclear if theoretical results reflect the scarce published data. Developing mathematical models for MEOR is very challenging since physical, chemical and biological factors need to be considered.

Published MEOR models are composed of transport properties, conservation laws, local equilibrium, breakdown of filtration theory and physical straining. Such models are so far simplistic and they were developed based on:

A Fundamental conservation laws, cellular growth, retention kinetics of biomass, and biomass in oil and aqueous phases. The main aim was to predict porosity retention as a function of distance and time.

B Filtration model to express bacterial transport as a function of pore size; and relate permeability with the rate of microbial penetration by applying Darcy's law.

Chemical kinetics is fundamental for coupling bioproduct formation to fluxes of aqueous species and suspended microbes. Fully numerical approaches have also been followed. For instance, coupled nonlinear parabolic differential equations: adding equation for the rate of diffusion of microbes and their capture by porous medium; differential balance equations for nutrient transport, including the effect of adsorption; and the assumption of bacterial growth kinetic based on Monod equation. Monod equation is indiscriminately used in modelling software, and has a limited behaviour which is inconsistent with the law of mass action that form the basis of kinetic characterisation of microbial growth. Application of law of mass action to microbial populations results in the linear logistic equation. And the application of the law of mass action to an enzyme-catalysed process results in the Michaelis-Menten equation, from which Monod is inspired. This makes things difficult for in situ biosurfactant production because controlled experimentation is required to determine specific growth rate and Michaelis-Menten parameters of rate-limiting enzyme reaction.

Modelling of bioclogging is complicated because the production of clogging metabolite is coupled nonlinearly to the growth of microbes and flux of nutrients transported in the fluid. Published models disregard the ecophysiology of the entire microbial microcosms at oil reservoir conditions. Microorganisms are a kind of catalyst whose activity (physiology) depends on the mutual interplay with other microbes and the environment (ecology). In nature, living and non living elements interact with each other in a complicated network of nutrients and energy. Some microbes produce extracellular polymeric substances and therefore its behaviour in pours media needs to consider both occupation by the EPS and the microbes themselves. Knowledge is lacking in this respect and therefore the aim of maximising yield and minimising cost remains unachieved.

Realistic models for MEOR at the conditions of the oil reservoir are missing, and reported parallel-pore models had fundamental deficiencies that were overcome by models considering the clogging of pores by microbes or biofilms, but such models have also the deficiency of being two-dimensional. The utilisation of such models in three dimensional models has not been proven. It is uncertain if they can be incorporated to popular oilfield simulation software. Thus, a field strategy needs a simulator capable of predicting bacterial growth and transport through porous network and in situ production of MEOR agents.

Chapter 7

Oil Platform

An oil platform, also referred to as an offshore platform or, somewhat incorrectly, oil rig, is a làrge structure with facilities to drill wells, to extract and process oil and natural gas, and to temporarily store product until it can be brought to shore for refining and marketing. In many cases, the platform contains facilities to house the workforce as well. Depending on the circumstances, the platform may be fixed to the ocean floor, may consist of an artificial island, or may float.

Remote subsea wells may also be connected to a platform by flow lines and by umbilical connections; these subsea solutions may consist of single wells or of a manifold centre for multiple wells.

History

Around 1891 the first submerged oil wells were drilled from platforms built on piles in the fresh waters of the Grand Lake St. Marys (a.k.a. Mercer County Reservoir) in Ohio. The wide but shallow reservoir was built from 1837 to 1845 to provide water to the Miami and Erie Canal. Around 1896 the first submerged oil wells in salt water were drilled in the portion of the Summerland field extending under the Santa Barbara Channel in California. The wells were drilled from piers extending from land out into the channel.

Other notable early submerged drilling activities occurred on the Canadian side of Lake Erie in the 1900s and Caddo Lake in Louisiana in the 1910s. Shortly thereafter, wells were drilled in tidal zones along the Gulf Coast of Texas and Louisiana. The Goose Creek field near Baytown, Texas is one such example. In the 1920s drilling was done from concrete platforms in Lake Maracaibo, Venezuela.

The oldest subsea well recorded in Infield's offshore database is the Bibi Eibat well which came on stream in 1923 in Azerbaijan. Landfill was used to raise shallow portions of the Caspian Sea.

In the early 1930s the Texas Company developed the first mobile steel barges for drilling in the brackish coastal areas of the gulf.

In 1937 Pure Oil Company (now part of Chevron Corporation) and its partner Superior Oil Company (now part of ExxonMobil Corporation) used a fixed platform to develop a field in 14 feet (4.3 m) of water, one mile (1.6 km) offshore of Calcasieu Parish, Louisiana.

In 1946, Magnolia Petroleum Company (now part of ExxonMobil) erected a drilling platform in 18 ft (5.5 m) of water, 18 miles off the coast of St. Mary Parish, Louisiana.

In early 1947 Superior Oil erected a drilling/production platform in 20 ft (6.1 m) of water some 18 miles off Vermilion Parish, Louisiana. But it was Kerr-McGee Oil Industries (now Anadarko Petroleum Corporation), as operator for partners Phillips Petroleum (ConocoPhillips) and Stanolind Oil & Gas (BP), that completed its historic Ship Shoal Block 32 well in October 1947, months before Superior actually drilled a discovery from their Vermilion platform farther offshore. In any case, that made Kerr-McGee's well the first oil discovery drilled out of sight of land.

The Thames Sea Forts of World War II are considered the direct predecessors of modern offshore platforms. Having been pre-constructed in a very short time, they were then floated to their location and placed on the shallow bottom of the Thames estuary.

Types

Larger lake- and sea-based offshore platforms and drilling rigs are some of the largest moveable man-made structures in the world. There are several types of oil platforms and rigs:

Fixed Platform

A fixed platform is a type of offshore platform used for the production of oil or gas. These platforms are built on concrete and/or steel legs anchored directly onto the seabed, supporting a deck with space for drilling rigs, production facilities and crew quarters. Such platforms are, by virtue of their immobility, designed for very long term use (for instance the Hibernia platform). Various types of structure are used, steel jacket, concrete caisson, floating steel and even floating concrete. Steel jackets are vertical sections made of tubular steel members, and are usually piled into the seabed. Concrete caisson structures, pioneered by the Condeep concept, often have in-built oil storage in tanks below the sea surface and these tanks were often used as a flotation capability, allowing them to be built close to shore (Norwegian fjords and Scottish firths are popular because they are sheltered and deep enough) and then floated to their final position where they are sunk to the seabed. Fixed platforms are economically feasible for installation in water depths up to about 1,700 feet (520 m).

Compliant Tower

A compliant tower (CT) is a fixed rig structure normally used for the offshore production of oil or gas. The rig consists of narrow, flexible (*compliant*) towers and a piled foundation supporting a conventional deck for drilling and production operations. Compliant towers are designed to sustain significant lateral deflections and forces, and are typically used in water depths ranging from 1,500 and 3,000 feet (450 and 900 m). At present the deepest is Baldpate in 580 m of water.

With the use of flex elements such as flex legs or axial tubes, resonance is reduced and wave forces are de-amplified. This type of rig structure can be configured to adapt to existing fabrication and installation equipment. Compared with floating systems, such as Tension-leg platforms and SPARs, the production risers are conventional and are subjected to less structural demands and flexing. This flexibility allows it to operate in much deeper water, as it can 'absorb' much of the pressure exerted on it by the wind and sea. Despite its flexibility, the compliant tower system is strong enough to withstand hurricane conditions. The first tower emerged in the early 1980s with the installation of Exxon's Lena oil platform.

Jackup Rig

A jack-up rig is a type of mobile platform that is able to stand still on the sea floor, resting on a number of supporting legs. The most popular designs use 3 independent legs, although some jackups have 4 legs or more. On "mat-type" jackups the legs are connected to a submerged hull.

Operation

A jackup is a floating barge fitted with long support legs that can be raised or lowered. The jackup is maneuvered (self-propelled or by towing) into location with its legs up and the hull floating on the water. Upon arrival at the work location, the legs are jacked down onto the seafloor. Then "preloading" takes place, where the weight of the barge and additional ballast water are used to drive the legs securely into the seabottom so they will not penetrate further while operations are carried out. After preloading, the jacking system is used to raise the entire barge above the water to a predetermined height or "air gap", so that wave, tidal and current loading acts only on the relatively slender legs and not on the barge hull.

Modern jacking systems use a rack and pinion gear arrangement where the pinion gears are driven by hydraulic or electric motors and the rack is affixed to the legs. Jackup rigs can only be placed in relatively shallow waters, generally less than 400 feet (120 m) of water. However, a specialised class of jackup rigs known as premium or ultra-premium jackups are known to have operational capability in water depths ranging from 500 to 625 feet.

Types

Drilling Platforms: This type of rig is commonly used in connection with oil and/or natural gas drilling. There are more jackup rigs in the worldwide offshore rig fleet than any other type of mobile offshore drilling rig. Other types of offshore rigs include semi-submersibles (which float on pontoon-like structures) and drillships, which are ship-shape vessels with rigs mounted in the centres. These rigs drill through holes in the drillship hulls, known as moon pools.

Barges: Jackup rigs also refer to specialised barges that are similar to an oil and gas platform but are used as a base for servicing other structures such as offshore wind turbines, long bridges, and drilling platforms.

Drillship

A drillship is a maritime vessel that has been fitted with drilling apparatus. It is most often used for exploratory offshore drilling of new oil or gas wells in deep water or for scientific drilling. The drillship can also be used as a platform to carry out well maintenance or completion work such as casing and tubing installation or subsea tree installations. It is often built to the design specification of the oil production company and/or investors, but can also be a modified tanker hull outfitted with a dynamic positioning system to maintain its position over the well. The greatest advantages these modern drillships have is their ability to drill in water depths of more than 2500 metres and the time saved sailing between oilfields worldwide. Drillships are completely independent, in contrast to semi-submersibles and jackup barges.

In order to drill, a marine riser is lowered from the drillship to the seabed with a blowout preventer (BOP) at the bottom that connects to the wellhead.

Drillships are just one way to perform exploratory drilling. This function can also be performed by semi-submersibles, jackup barges, barges, or platform rigs.

The first drillship was the Cuss 1, and the fleet size has been growing ever since. By 2013 the worldwide fleet of drillships is expected to top 80 ships, more than double its size in 2009. Drillships are not only growing in size but also in capability with new technology assisting operations from academic research to ice drilling. U.S. President Barack Obama's decision in late March 2010 to expand U.S. domestic exploratory drilling seems likely to increase further developments of drillship technology.

Stena drillings 'DrillMAX ICE' being constructed by Samsung Heavy Industries in South Korea is to be the most expensive drillship built to date and is to complete exploration in Arctic regions when completed in 2012.

Floating Production Storage and Offloading

A floating production, storage and offloading (FPSO) unit is a floating vessel used by the offshore industry for the processing of hydrocarbons and for storage of oil. A FPSO vessel is designed to receive hydrocarbons produced from nearby platforms or subsea template, process them, and store oil until it can be offloaded onto a tanker or transported through a pipeline. FPSOs are preferred in frontier offshore regions as they are easy to install, and do not require a local pipeline infrastructure to export oil. FPSOs can be a conversion of an oil tanker or can be a vessel built specially for the application. A vessel used only to store oil (without processing it) is referred to as a floating storage and offloading vessel (FSO).

History

Oil has been produced from offshore locations since the late 1940s. Originally, all oil platforms sat on the seabed, but as exploration moved to deeper waters and more distant locations in the 1970s, floating production systems came to be used.

The first oil FPSO was the *Shell Castellon*, built in Spain in 1977.

The *Sanha* LPG FPSO operates offshore Angola, and is the first such vessel with complete onboard liquefied petroleum gas processing and export facilities. It can store up to 135,000 cubic metres of LPG while awaiting export tankers for offloading. In the opposite (discharge and regasification) end of the LNG chain, the first ever conversion of an LNG carrier (Golar LNG owned Moss type LNG carrier) into an

LNG floating storage and regasification unit was carried out in 2007 by Keppel shipyard in Singapore. An LNG FPSO works under the same principles an oil FPSO works under, taking the well stream and separating out the natural gas (primarily methane and ethane) and producing LNG, which is stored and offloaded. On July 29, 2009, Shell and Samsung announced an agreement to build up to 10 LNG FPSOs: Already Flex LNG has four contracts for smaller units at the same yard.

On May 20 2011, Royal Dutch Shell announced the planned development of a Floating Liquefied Natural Gas (FLNG) facility, which will be situated 200km off the coast of Western Australia and is due for completion in around 2017. When it is finished, this will be the largest floating offshore facility. It will measure around 488m long and 74m wide, and when fully ballasted will weigh 600,000 tonnes.. It will have a total storage capacity of 436,000 cubic metres of LNG, plus LPG condensate.

Mechanisms

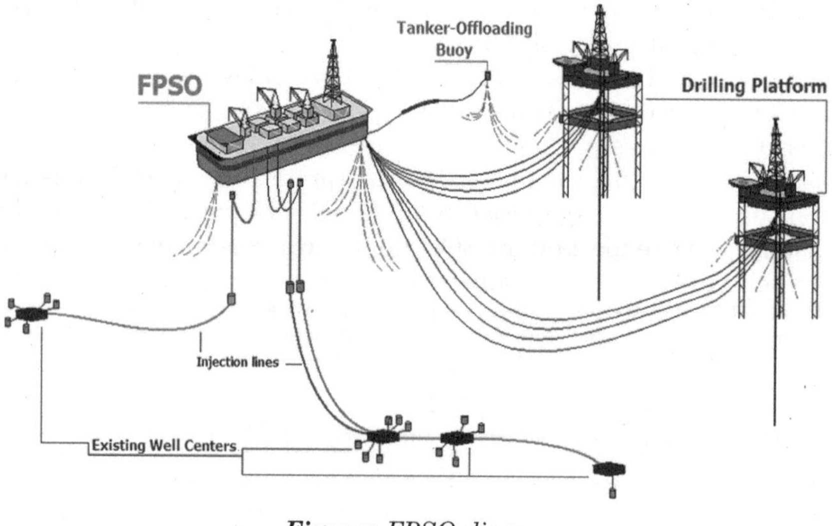

Figure: *FPSO diagram*

Oil produced from offshore production platforms can be transported to the mainland either by pipeline or by tanker. When a tanker is chosen to transport the oil, it is necessary to accumulate oil in some form of storage tank such that the oil tanker is not continuously occupied during oil production, and is only needed once sufficient oil has been produced to fill the tanker. At this point the transport tanker connects to the stern of the storage unit and offloads oil.

In the early days, the storage units consisted of decommissioned oil tankers, which were stripped down and equipped with process/ production facilities (becoming therefore FPSOs), and were connected to a permanent mooring point. Today, there are two main types of FPSOs, those built converting an existing oil tanker, and those that are purpose-built. The FPSO design will depend on the area of operation. In benign waters the FPSO may have a simple box shape or it may be a converted tanker. Generally (but not always) the production lines (risers) are connected to a major component of the vessel, called a Turret, which allows the vessel to rotate in order to head into the wind and reduce environmental forces on the moorings. In relatively calm waters, such as in West Africa, turrets can be located externally to the ship structure, hanging off the bow of the FPSO. For harsher environments like the North Sea, the turret is generally located internally. The turrets and the mooring systems can be designed to be disconnectible or to remain permanently moored. Most ship-shaped FPSOs in the North Sea are purpose-built and are permanently moored.

While most FPSOs are ship-shaped, some FPSOs have a semi-submersible type hull with storage (very rare), or have a cylindrical hull. The inherent symmetry of these FPSO configurations makes turrets unnecessary, so the platforms remain in a fixed orientation.

An FPSO has the capability to carry out some form of separation process. If the unit does not have such facilities, it is generally referred to as a Floating Storage and Offloading unit, and would be operated in conjunction with a production platform. Process plant on FPSO is a core component on facility and forms a key part of production process. Production is usually conducted in 3 phases 1 Separation of Gas. 2 Separation of Water. 3 Separation of oil. Gas recovered/separated during production may be used as fuel on Marine energy resource units (MRU) fitted on board. Gas may be flared off in some cases if MRU is not fitted. Water separation may be carried out using Dehydrators or Hydro Cyclones

Advantages

Floating production, storage and offloading vessels are particularly effective in remote or deepwater locations where seabed pipelines are not cost effective. FPSOs eliminate the need to lay expensive long-distance pipelines from the oil well to an onshore terminal. They can also be used economically in smaller oil fields which can be exhausted in a few years and do not justify the expense of installing a pipeline. Once the field is depleted, the FPSO can be moved to a new location. In areas of the world subject to cyclones (northwestern Australia) or icebergs (Canada), some FPSOs are able to release their mooring/riser turret and steam away to safety in an emergency. The turret sinks beneath the waves and can be reconnected later.

Specific Types

A floating storage and offloading unit (FSO) is a floating storage device, which is a simplified FPSO without the capability for oil or gas processing. Most FSOs are old single hull supertankers that have been converted. An example is *Knock Nevis, ex Seawise Giant*, was for long time world's largest ship, which had been converted to an FSO to be used offshore Qatar. At the other end of the LNG logistics chain, where the natural gas is brought back to ambient temperature and pressure, ships may also be used as FSRUs. A LNG floating storage and regasification unit (FSRU) is a floating storage and regasification system, which receives liquefied natural gas (LNG) from offloading LNG carriers, and the onboard regasification system provides natural gas send-out through flexible risers and pipeline to shore. Mooring systems for FSO, FPSO & FSU units are available in market which allow the vessel to be moored on a ice sheet. Ice sheet used for mooring this case may be located below the ice sheet under water.

Vessels

Records: The FPSO operating in the deepest water depth is the FPSO BW Pioneer from BW Offshore operated on behalf of Petrobras Americas INC.. The FPSO is moored at a depth of 2,600 m in Block 249 Walker Ridge in the US GoM and is rated for 100,000 bbl/d (16,000 m^3/d). The EPCI contract was awarded in October 2007 and first oil is planned for 3rd Q 2011. The FPSO conversions were done at Keppel Shipyard Tuas in Singapore and the topsides were fabricated in modules at various international vendors. The FPSO has a disconnectable turret (APL). The vessel can disconnect for hurricanes and reconnect with minimal down time. The world's largest FPSO is the *Kizomba A*, with a storage capacity of 2.2 million barrels (350,000 m^3). Built at a cost of over US$800 million by Hyundai Heavy Industries in Ulsan, Korea, it is operated by Esso Exploration Angola (ExxonMobil). Located in 1200 metres (3,940 ft) of water at Deepwater block 200 statute miles (320 km) offshore in the Atlantic Ocean from Angola, Central Africa, it weighs 81,000 tonnes and is 285 metres long, 63 metres wide, and 32 metres high (935 ft by 207 ft (63 m) by 105 ft). The first FSO in the Gulf of Mexico, The FSO Ta'Kuntah, has been in operation since August 1998. The FSO is under a service agreement with PEMEX Exploration and Production. MODEC owns and operates the FSO. The FSO Ta'Kuntah was installed as part of the Cantarell

Field Development in the Gulf of Mexico. The field is located in the Bay of Campeche, offshore Mexico's Yucatan peninsula. The FSO Ta'Kuntah is a converted ULCC tanker with a SOFEC external turret mooring system, two flexible risers connected in a lazy-S configuration between the turret and a pipeline end manifold (PLEM) on the seabed, and a unique offloading system. The FSO is designed to handle 800,000 bbl/d (130,000 m^3/d) with no allowance for downtime.

The world's smallest FPSO is the Crystal Ocean, operating in 137 m of water in the Bass Strait between Australia and Tasmania on the Basker Manta Field. It is leased by Roc Oil (Sydney-based international petroleum exploration and production company) from Rubicon Offshore and is operated on their behalf by AGR Asia Pacific; it is currently producing 5,000 bbl/d (790 m^3/d).

The FPSO in the shallowest water depth of just 13 m is the Armada Perkasa in the Okoro field in Nigeria, West Africa, for Afren Energy. This spread moored (fixed orientation) vessel uses 100 mm, 150 mm and 200 mm bore DeepFlex non-steel flexible risers in a double lazy wave formation (with weights and distributed buoyancy) to accommodate the large motion offsets in an environment of extreme waves and currents. The Skarv FPSO, developed and engineered by Aker Solutions for BP Norge, will be the most advanced and largest FPSO deployed in the Norwegian Sea, offshore Mid Norway. Skarv is a gas condensate and oil field development. The development will tie in five sub-sea templates, and the FPSO has capacity to include several smaller wells nearby in the future. The process plant on the vessel can handle about 19,000,000 cubic metres per day (670,000,000 cu ft/d) of gas and 13,500 cubic metres per day (480,000 cu ft/d). An 80 km gas export pipe will tie in to Åsgard transport system. Aker Solutions (formerly Aker Kvaerner) developed the front-end design for the new floating production facility as well as the overall system design for the field and preparation for procurement and project management of the total field development. The hull is an Aker Solutions proprietary "Tentech™975" design. BP also selected Aker Solutions to perform the detail engineering, procurement and construction management assistance (EPcma) for the Skarv field development. The EPcma contract covers detail engineering and procurement work for the FPSO topsides as well as construction management assistance to BP including hull and topside facilities. The production start for the field is scheduled for August 2011. BP awarded the contract for fabrication of the Skarv FPSO hull to Samsung Heavy Industries in South Korea and the Turret contract to SBM. The FPSO has a length of 292m, breadth of 50.6m and is 29m deep and accommodate 100 people in single cabins. The hull will be delivered in January 2010.

In July 2011, Shell announced the world's largest FPSO, scheduled to come on stream in 2017.

Tension-leg Platform

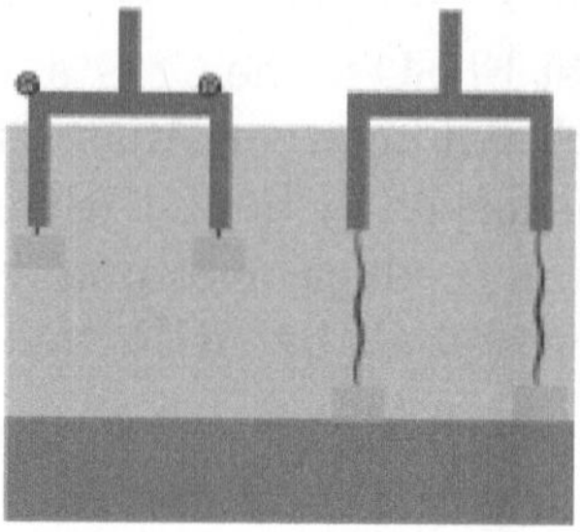

Figure: *A tension-leg platform (grey) under tow with seabed anchors (light grey) held up by cables (red) on left-hand side; platform with seabed anchors lowered and cables lightly tensioned on right-hand side*

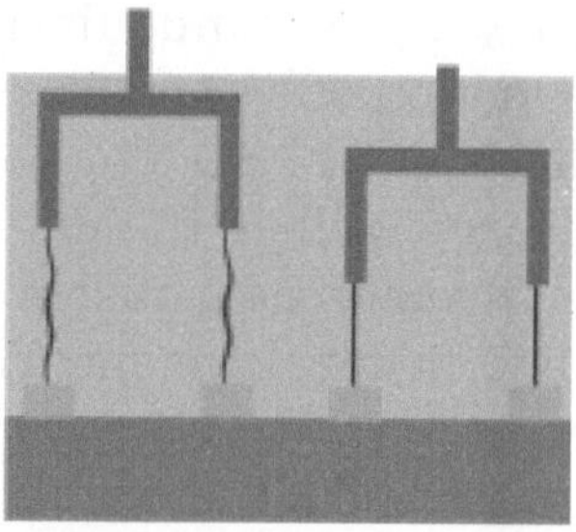

Figure: *Tension leg platform (grey) free floating on left-hand side; structure is pulled by the tensioned cables (red) down towards the seabed anchors (light-grey) on right-hand side (very simplified, omitting details of temporary ballast transfers)*

A Tension-leg platform or Extended Tension Leg Platform (ETLP) is a vertically moored floating structure normally used for the offshore production of oil or gas, and is particularly suited for water depths greater than 300 metres (about 1000 ft) and less than 1500 metres (about 4900 ft). Use of tension-leg platforms has also been proposed for wind turbines. The platform is permanently moored by means of tethers or tendons grouped at each of the structure's corners. A group of tethers is called a tension leg. A feature of the design of the tethers is that they have relatively high axial stiffness (low elasticity), such that virtually all vertical motion of the platform is eliminated. This allows the platform to have the production wellheads on deck (connected directly to the subsea wells by rigid risers), instead of on the seafloor.

This allows a simpler well completion and gives better control over the production from the oil or gas reservoir, and easier access for downhole intervention operations.

TLP's have been in use since the early 1980s. The first Tension Leg Platform was built for Conoco's Hutton field in the North Sea in the early 1980s. The hull was built in the dry-dock at Highland Fabricator's Nigg yard in the north of Scotland, with the deck section built nearby at McDermott's yard at Ardersier. The two parts were mated in the Moray Firth in 1984.

Larger TLP's will normally have a full drilling rig on the platform with which to drill and intervene on the wells. The smaller TLPs may have a workover rig, or in a few cases no production wellheads located on the platform at all. The deepest (E)TLPs measured from the sea floor to the surface are:

- 4,674 ft (1,425 m) Magnolia ETLP. Its total height is some 5,000 feet (1,500 m).
- 4,300 ft (1,300 m) Marco Polo TLP
- 4,250 ft (1,300 m) Neptune TLP
- 3,863 ft (1,177 m) Kizomba A TLP
- 3,800 ft (1,200 m) Ursa TLP. Its height above surface is 485 ft (148 m) making a total height of 4,285 ft (1,306 m).
- 3,350 ft (1,020 m) Allegheny TLP
- 3,300 ft (1,000 m) W. Seno A TLP

Use for Wind Turbines

The Massachusetts Institute of Technology and the National Renewable Energy Laboratory first explored the concept of TLPs for offshore wind turbines in September 2006. Earlier offshore wind turbines cost more to produce, stood on towers dug deep into the ocean floor, were only possible in depths of at most 50 feet (15 m), and generated 1.5 megawatts for onshore units and 3.5 megawatts for conventional offshore setups. In contrast, TLP installation was calculated to cost a third as much. TLPs float, and researchers estimate they can operate in depths between 100 and 650 feet (200 m) and farther away from land, and they can generate 5.0 megawatts.

TLPs cost less to make and install because they are assembled onshore, are towed to their destination, and can be moved. Paul Sclavounos, an MIT professor of mechanical engineering and naval architecture who was involved in the design, said, "You don't pay anything to be buoyant."

Computer simulations project that in a hurricane TLPs would shift three to six feet and the turbine blades would cycle above wave peaks. MIT and NREL researchers say dampers could be used to reduce motion in the event of a natural disaster. MIT and NREL researchers plan to install a half-scale prototype south of Cape Cod. Sclavounos said, "We'd have a little unit sitting out there to show that this thing can float and behave the way we're saying it will."

Gravity-Based Structure

A gravity-based structure (GBS) is a support structure held in place by gravity. A common application for a GBS is an offshore oil platform. These structures are often constructed in fjords since their protected area and sufficient depth are very desirable for construction. A GBS intended for use as an offshore oil platform is constructed of steel reinforced concrete, often with tanks or cells which can be used to control the buoyancy of the finished GBS. When completed, a GBS is towed to its intended location and sunk. The platform structure which a GBS supports is called the topsides.

Gravity-based structures are also widely used for offshore wind power plants. By the end of 2010, 14 of the world's offshore wind farms were supported by Gravity-based structures. The GBS are suited for water depths up to 20 metres. The deepest registered offshore wind farm with Gravity-based structures is Thornton Bank 1, Belgium, with a depth up to 27.5 metres. However, as offshore wind power plants are growing in size and moving towards deeper waters, the GBS is not considered competitive in comparison with other support structures.

Spar (Platform)

A spar, named for logs used as buoys in shipping and moored in place vertically, is a type of floating oil platform typically used in very deep waters. Spar production platforms have been developed as an alternative to conventional platforms. A spar platform consists of a large-diametre, single vertical cylinder supporting a deck. The first spar platform in the Gulf of Mexico was installed in September 1996.

The world's deepest platform is the Perdido in the Gulf of Mexico, floating in 2,438 metres of water. It is operated by Royal Dutch Shell and was built at a cost of $3 billion.

Normally Unmanned Installation

A Normally Unmanned Installation (NUI) is a type of automated offshore Oil/Gas platform designed to be primarily operated remotely, without the constant presence of personnel. These generally were characterised by their small size, often consisting of just a well bay with a helipad on top. They are often a compromise of providing the convenience of surface wellheads, which are easier to build and maintain, while avoiding the high operating costs of a full production platform. They are generally only used in shallower water, where constructing many small NUIs is a relatively easy and cheap option as compared to the cost of using subsea wells.

This can be seen in the Southern North Sea where large numbers of wells are on smaller NUIs, compared with the more northern areas of the continental shelf where fewer larger platforms and subsea sites are the norm. NUIs are commonly serviced from a nearby larger platform, e.g., Mungo serviced from Marnock. These installations will include an emergency shelter with essential food and water in order to provide a safe refuge in the event that weather or other considerations prevent a visiting crew from returning to base. Regular visits may be made for routine maintenance and for smaller well work such as wireline operations. Anything larger requires a drilling rig to be brought in, but this is still an advantage over subsea wells, which require a drilling rig or light intervention vessel for any well intervention.

In recent times NUI has become a phase within the decommissioning of previously manned offshore installations. Platforms would generally be positively isolated from hydrocarbons and flushed clean to suit environmental issues. Platforms would then only be visited infrequently for integrity checks and maintenance of temporary equipment left for power requirements and safety functionality. The platform would then remain in this phase until further decommissioning activities are carried out to remove the topsides in modules or piece small, then remove the jacket.

Chapter 8

Expandable Tubular Technology

Expandable tubular technology is a system for increasing the diametre of the casing or liner of an oil well by up to 30% after it has been run down-hole.

Challenges

Two challenges facing the Oil & Gas industry are accessing new reservoirs that currently cannot be reached economically and maintaining profitable production from producing older fields. Expandable tubular technology is considered one of the most revolutionary technologies that has benefited the oil and gas industry during the last 10 years. The development of this technology was initiated by the industry need to reduce drilling costs, increase production of tubing constrained wells and to enable operators to access reservoirs that could otherwise not be reached economically. Expanded casing applications concentrate on reducing the telescopic profile of well designs through a downhole tube expansion process. Wider applications of the technology exist for example water shut off, and casing repairs in old wells.

Telescoping Well Design

To understand the enormous potential value of expandable tubular technology one must first understand the concept of telescoping well design and the constraints it places on conventional drilling.

During the planning stages, one of the primary considerations that the well engineer takes into account is the density (or "weight") of the drilling fluid. It must be heavy enough to suppress any pressured fluids contained by the rock (thus avoiding a blowout), and light enough to avoid breaking down the rock itself. Both parameters generally increase with depth, and create a window between which the drilling fluid density is safe. However, the mud density required to suppress fluid pressures at 9000' would usually break down the formation further up the well at 1000'. Therefore, the well is drilled in sections by running casing strings to cover depth ranges between which the required mud densities are suitable for the entire range.

However, each bit must be smaller than the previous casing string, which in turn has to be smaller than the previous hole.

This requirement of well design creates a well where each hole section is ever-decreasing in diametre. The only conventional way to combat this effect is to start with an enormous hole at the top (sometimes 30") in order to run five casing strings and still end up with a 6" hole in the targeted reservoir. And this well design leaves little room for error: if the drillers are forced to run a casing string on account of unexpected formation pressures then the target formation may be impossible to reach.

Thus there is an enormous carrot in a casing system that loses minimal - or possibly no - diametre while still isolating a whole section. Many current systems have the aim of producing a monobore well, that is, a well with a single diametre from top to bottom. But the shallow wells can the drilled with uniform diametre.

Technology

Figure: *Image of Expanded Tubular FE Analysis*

To reduce the loss of diametre each time a new casing string or liner is set, a cold working process has been developed whereby the casing or liner can be expanded by up to 20% in diametre after being run down-hole.

For this purpose, an expansion tool that exceeds the inner diametre of the tube by the required amount of expansion is forced through the pipe. This is done either hydraulically, by applying mud pressure, or mechanically, by pulling the conical/tapered expansion tool. The expansion needs to be reliable, when expanding several thousand feet below the surface. This can be from 30–6,000 ft in length.

Applications

The applications can be groups into two main categories – Cased hole and Open hole. Cased hole work is mainly down during the work over or completion phase of a well. The open hole products are used during the drilling period of a well.

The products developed and available today in cased hole work are the expandable liner hanger and the cased hole clad. The expandable liner hanger is basically an evolution of existing equipment currently used in the oil industry, a product with better through bore and envisaged higher reliability. The Case hole clad provide a casing patch across a damaged section of casing, or to close off previously perforated casing. This product has two main advantages – minimal through bore loss [basically two times the wall thickness of tubular being expanded] and high pressure integrity performance.

Open hole applications is where expandable technology brings real advantages to the operator. Currently the products available are open hole liner and open hole clads.

Casing String

Casing string is a long section of connected oilfield pipe that is lowered into a wellbore and cemented. The pipe segments (called "joints") are typically about 40 feet (12 m) in length, male threaded on each end and connected with short lengths of double-female threaded pipe called couplings. (Some speciality casing is manufactured in one piece with a female thread machined directly into one end.)

Specification 5C3 of the American Petroleum Institute standardizes 14 casing sizes from 4.5 inches (11.43 cm) to 20 inches (50.80 cm) outside diametre ("OD"). This and related API documents also promulgate standards for the threaded end finish, the wall thickness (several are available in each size to satisfy various design parameters, and in fact are indirectly specified by standardized nominal weights per linear foot; thicker pipe obviously being heavier), and the strength and certain chemical characteristics of the steel material. Several material strengths—termed "Grades" and ranging from 40,000 pounds per square inch (276 MPa) to 150,000 pounds per square inch (1,034 MPa) minimum yield strength—are available for most combinations of OD and wall thickness to meet various design needs. Finally, the API publications provide performance minimums for longitudinal strength ("joint strength") as well as resistance to internal (bursting) and external (collapsing) pressure differentials.

A typical piece of casing might be described as 9-5/8" 53.5# P-110 LT&C Rg 3: specifying OD, weight per foot (53.5 lbm/ft thus 0.545-inch wall thickness and 8.535-inch inside diametre), steel strength (110,000 psi yield strength), end finish ("Long Threaded and Coupled"), and approximate length ("Range 3" usually runs between 40 and 42 feet).

Casing is run to protect or isolate formations adjacent to the wellbore. It is generally not possible to drill a well through all of the formations from surface (or the seabed) to the target depth in one hole section. For example, fresh-water-bearing zones (usually found only near the surface) must be protected soon after being penetrated. The well is therefore drilled in sections, with each section of the well being sealed off by lining the inside of the borehole with steel pipe, known as casing, and filling the annular space (or at least the lower portion) between this casing string and the borehole with cement. Then drilling commences on the subsequent hole section, necessarily with a smaller bit diametre that will pass through the newly installed casing.

A *liner* is a casing string that does not extend to the surface, being hung instead from a liner hanger set inside of the previous casing string but usually within about 300 feet (91 m) of its bottom. Other than the obvious cost savings, the liner installation allows larger drill pipe or production tubing to be used in the upper portions of the well. (A disadvantage is the occasional difficulty in effecting a pressure seal by squeeze cementing the casing-liner overlap zone.)

Depending on the conditions encountered (e.g., zones of differing formation pressure gradients), three or four casing strings may be required to reach the target depth. The cost of the casing can constitute 20-30% of the total cost of the well. Great care must therefore be taken when designing a casing programme that will meet the requirements of the well.

Plasma Deep Drilling Technology

In the recent past, there were several attempts to develop new drilling technology which would be able to substitute conventional contact-based rotary systems. These concepts were based on different principles such as water jet, hydrothermal spallation or laser. Worldwide there are several research teams which work continually on research and development of new deep drilling technology. Until now, no one from these attempts has been proved to be effective in extreme deep drilling conditions. One of them was also technology of plasma in several different variants.

High-Energetic Electrical Plasma

High-energetic electrical plasma is one of the most promising technologies which are currently being developed in deep drilling applications. It has lower energy efficiency than some of the other technologies, but has several other advantages. The most important are utilisation in water environment or producing boreholes with wide range of diametres. The research team from Slovakia is developing drilling concept based on electrical plasma principle suitable for several drilling applications. The core of the research is held in Research Centre for Deep Drilling established in the premises of Slovak Academy of Sciences in Bratislava.

Physical Principle of Electrical Plasma

An electric arc is an electrical breakdown of a gas which produces an ongoing plasma discharge, resulting from a current flowing through normally nonconductive media such as air or gas. An arc discharge is characterised by a lower voltage than a glow discharge, and relies on thermionic emission of electrons from the electrodes supporting the arc. The electric arc is influenced by different factors: the gas flow, inner and outer magnetic fields, construction elements of the chamber which confine the arc, the elements being often under electric voltage, differing from that of the arc. Thus, there is a strong need for deep understanding of a wide spectrum of the processes taking place in the discharge chamber in order to develop highly effective plasma torches that we have proposed to use as a source of the thermal plasma.

Plasma Deep Drilling Technology in Mining Applications

Plasma deep drilling technology may be suitable not only for geothermal, oil and gas or carbon capture and storage applications but also for underground mining. This new non-contact drilling method combined with the development of current mining techniques may bring smart technology for underground mining industry. The most promising project connects advantages of plasma drilling and raise boring technologies.

Advantages of Plasma Deep Drilling Technology

1. Higher drilling energy efficiency
2. Continuous drilling process without replacement of mechanical parts

3. Constant casing diametre
4. Effective transport of disintegrated rock

Cementing Equipment

Cementing Equipment includes various items used while drilling oil/gas/water wells. Casing will be installed at various depths while drilling. It is held in place by cement, which also provides zone isolation. Down-hole equipment such as Centralisers and Turbolizers is used to ensure obtaining satisfactory zone isolation. Float Shoes and Collars (float valves) prevent backflow after the cement has been pumped into place.

Centralisers, Turbolizers and Scratchers

Devices fitted with hinged collars and bow springs help keep the casing or liner in the centre of the wellbore to help ensure efficient placement of a cement sheath around the casing string. If casing strings are cemented off-centre, there is a high risk that a channel of drilling fluid or contaminated cement will be left where the casing contacts the formation, creating an imperfect seal.

Turbolizers have added fins to "stir" up the drilling fluid and cement to keep the flow turbulent in order to make sure the cement flows all the way around the casing in order to prevent channelling. Scratchers use metal wires to scrape mud cake off permeable zones to help obtain a solid cement column.

Float Shoe

A rounded profile float shoe with an integral check valve attached to the bottom of a casing string prevents reverse flow, or U-tubing, of cement slurry from the annulus into the casing or flow of wellbore fluids into the casing string as it is run. The float shoe also guides the casing toward the centre of the hole to minimise hitting rock ledges or washouts as the casing is run into the wellbore. By "floating" casing in, hook weight is reduced. With controlled or partial fill-up as the string is run, the casing string can be floated into position, precluding the need for the rig to carry the entire weight of the casing string. The outer portions of the float shoe are made of steel and generally match the casing size and threads, although not necessarily the casing grade. The inside (including the external taper) is usually made of cement or thermoplastic, since this material must be drilled out if the well is to be deepened beyond the casing point. Guide shoes are a variant of a float shoe without a check valve.

Float Collar

A float collar is installed near the bottom of the casing string. Cement plugs land on it during the primary cementing operation, thus retaining inside the casing a small portion of the cement slurry that may have become contaminated as the top plug scrapes the inside of the casing. It is similar to a float shoe; often both are used for redundancy. The internal check valves may be flapper type or spring-loaded balls. The check-valve assembly fixed within the float collar prevents flow back of the cement slurry when pumping is stopped. Without a float collar or float shoe, the cement slurry placed in the annulus could U-tube, or reverse flow back into the casing. The greater density of cement slurries than the displacement mud inside the casing causes the U-tube effect.

Stage Collar

Stage Collars provide a means for cement slurry to be displaced higher in the annulus immediately following the primary cement job. A series of differently sized rubber plugs pumped down inside the casing open and then later close the sliding valves.

Coiled Tubing Truck

A coiled tubing truck is used to deploy continuous tubing into an oil or gas well. They are normally used for well cleanouts (both over balanced or jetvac underbalanced, plug pulling or drill outs, fishing, spotting of acid or cement, installation of umbilicals such as flatak, or instrument and camera runs in vertical or horizontal oil and gas wells.

There are 4 types of coiled tubing trucks:

- Shallow coiled tubing trucks, sometimes called bobtail units or body load units. These units are used on wells that are less than 5,000 feet deep and usually are mounted on traditional single steering axle and tandom rear axle semi trucks. Some larger shallow units are mounted on tandem steering and tridem rear axle trucks. These units traditionally use 1" to 1.50" tubing and have 660 CFM air compressors mounted on the truck for self contained air to keep the well service as economic as possible. An example of a service provider of this would be Titan Oil and Gas Services.
- Intermediate Units are used on wells from 5,000 feet to 10,000 feet deep and mounted on semi truck and trailer units. These units usually use 1.75" or 2" for imtermediate well depths and if they are going deeper they will use 1.25".

- Large Coil units are used on wells 10,000 feet and deeper or used on intermediate well depths when the work calls high flows of nitrogen for large size tubing such as 2", 2 3/8" or 2 7/8".
- Portable Drilling Units. Drilling with coil tubing is a relatively new technology. The advantage of drilling with coil is the speed at which one can drill. There are no pipe connections to be made and as such the drilling can be accomplished very rapidly. The deepest coil drilling on land record currently is held by Extreme Drilling at 11,455 feet deep.

Conductor Pipe

The Conductor Pipe is a large diametre pile that is set into the ground to provide the initial stable structural foundation for a borehole or oil well. It can also be referred to as a *drive pipe* because it is often *driven* into the ground with a pile driver. Conductor pipe is tyically set, on petroleum wells, before any drilling operations are performed. It is usually set with special pile-driving or spudder rigs, though the drilling rig is sometimes used to save time and money. In the offshore drilling industry, the conductor pipe is set in the seabed, and is a key structural foundation for the subsea wellhead. When drilling an offshore well, a marine riser is connected to the well, and this is subjected to large environmental forces such as current and waves which are transferred to the conductor. The forces can be large and also cyclical which can potentially causing fatigue damage to conductor especially in the harsh environmental conditions. A Conductor tensioner unit (CTU) can be used in shallow water on jackup rig to withstand such damaging forces.

Drilling Riser

Figure: *Drilling riser joints with buoyancy modules*

A drilling riser is a conduit that provides a temporary extension of a subsea oil well to a surface drilling facility. Drilling risers are categorised into two types: marine drilling risers used with subsea blowout preventer (BOP) and generally used by floating drilling vessels; and tie-back drilling risers used with a surface BOP and generally deployed from fixed platforms or very stable floating platforms like a spar or tension leg platform (TLP).

Marine Drilling Riser

A marine drilling riser has a large diametre, low pressure main tube with external auxiliary lines that include high pressure choke and kill lines for circulating fluids to the subsea blowout preventer (BOP), and usually power and control lines for the BOP. The design and operation of marine drilling risers is complex, and the requirement for high reliability means an extensive amount of engineering analysis is required. When used in water depths greater than about 20 metres, the marine drilling riser has to be tensioned to maintain stability. A marine riser tensioner located on the drilling platform provides a near constant tension force adequate to maintain the stability of the riser in the offshore environment. The level of tension required is related to the weight of the riser equipment, the buoyancy of the riser, the forces from waves and currents, the weight of the internal fluids, and an adequate allowance for equipment failures.

To reduce the amount of tension required to maintain stability of the riser, buoyancy modules are added to the riser joints to make them close to neutrally buoyant when submerged. The international standard ISO 13624-1:2009 covers the design, selection, operation and maintenance of marine riser systems for floating drilling operations. Its purpose is to serve as a reference for designers, for those who select system components, and for those who use and maintain this equipment. It relies on basic engineering principles and the accumulated experience of offshore operators, contractors, and manufacturers.

Tie-Back Drilling Riser

A tie-back riser can be either a single large-diametre high pressure pipe, or a set of concentric pipes extending the casing strings in the well up to a surface BOP.

Hydraulic Fracturing

Hydraulic fracturing is the propagation of fractures in a rock layer caused by the presence of a pressurized fluid. Hydraulic fractures may form naturally, as in the case of veins or dikes, or may be man-made in order to release petroleum, natural gas, coal seam gas, or other substances for extraction, where the technique is often called fracking or hydrofracing. This type of fracturing, known colloquially as a frack job (or frac job), is done from a wellbore drilled into reservoir rock formations. The energy from the injection of a highly-pressurized fracking fluid, creates new channels in the rock which can increase the extraction rates and ultimate recovery of fossil fuels. The fracture width is typically maintained after the injection by introducing a *proppant* into the injected fluid. Proppant is a material, such as grains of sand, ceramic, or other particulates, that prevent the fractures from closing when the injection is stopped.

The practice of hydraulic fracturing has come under scrutiny internationally due to apparent concerns about the environment, health and safety, and has been suspended or banned in some countries.

Mechanics

Fracturing in rocks at depth is suppressed by the confining pressure, particularly in the case of tensile (Mode 1) fractures which require the walls of the fracture to move apart. Hydraulic fracturing occurs when the effective stress is reduced sufficiently by an increase in the pressure of fluids in the rock such that the minimum principal stress becomes tensile and exceeds the tensile strength of the material, allowing such fracture to occur. Fracture formed in this way will be orientated perpendicular to the minimum principal stress and for this reason, induced hydraulic fractures in wellbores are sometimes used to determine stress orientations. In natural examples, such as dikes or vein-filled fractures, their orientations can be used to infer past stress states.

Natural Examples

Rocks often contain evidence of past hydraulic fracturing events, where fluids have passed through tensile fractures.

Veins

Most vein systems are a result of repeated hydraulic fracturing during periods of relatively high pore fluid pressure. This is particularly clear in the case of 'crack-seal' veins, where the vein material can be seen to have been added in a series of discrete fracturing events, with extra vein material deposited on each occasion. One mechanism to explain such examples of long-lasting repeated fracturing, is the effects of seismic activity, in which the stress levels rise and fall episodically and large volumes of fluid may be expelled from fluid-filled fractures during earthquakes, a process referred to as 'seismic pumping'.

Dikes

High-level minor intrusions such as dikes propagate through the crust in the form of fluid-filled cracks, although in this case the fluid is magma. In sedimentary rocks with a significant water content the fluid at the propagating fracture tip will be steam.

Induced Hydraulic Fracturing

The technique of hydraulic fracturing is used to increase or restore the rate at which fluids, such as oil or water, or natural gas can be produced from subterranean natural reservoirs, including unconventional reservoirs such as shale rock or coal beds. Hydraulic fracturing enables the production of natural gas and oil from rock formations deep below the earth's surface (generally 5,000-20,000 feet or 1,500-6,100 m). At such depth, there may not be sufficient porosity and permeability to allow natural gas and oil to flow from the rock into the wellbore at economic rates. Thus, creating conductive fractures in the rock is essential to extract gas from shale reservoirs because of the extremely low natural permeability of shale, which is measured in the microdarcy to nanodarcy range. Fractures provide a conductive path connecting a larger area of the reservoir to the well, thereby increasing the area from which natural gas and liquids can be recovered from the targeted formation. While the main industrial use of hydraulic fracturing is in stimulating production from oil and gas wells, hydraulic fracturing is also applied to:

- Stimulating groundwater wells
- Preconditioning rock for caving or inducing rock to cave in mining
- As a means of enhancing waste remediation processes, usually hydrocarbon waste or spills
- Dispose of waste by injection into deep rock formations
- As a method to measure the stress in the earth
- For heat extraction to produce electricity in an Enhanced Geothermal System.

Method

A hydraulic fracture is formed by pumping the fracturing fluid into the wellbore at a rate sufficient to increase pressure downhole to exceed that of the fracture gradient of the rock. The rock cracks and the fracture fluid continues farther into the rock, extending the crack still farther, and so on. To keep this fracture open after the injection stops, a solid proppant, commonly a sieved round sand, is added to the fluid. The propped fracture is permeable enough to allow the flow of formation fluids to the well. Formation fluids include gas, oil, salt water, fresh water and fluids introduced to the formation during completion of the well during fracturing.

The location of fracturing along the length of the borehole can be controlled by inserting composite plugs, also known as bridge plugs, above and below the region to be fractured. This allows a borehole to be progressively fractured along the length of the bore without fluid leaking through previously fractured regions. Fluid and proppant are introduced to the working region through piping in the upper plug. This method is commonly referred to as "plug and perf." Typically, hydraulic fracturing is performed in cased wellbores and the zones to be fractured are accessed by perforating the casing at those locations.

Well Types

While hydraulic fracturing is many times performed in vertical wells, today it is more often performed in horizontal wells. Horizontal drilling involves wellbores where the terminal drillhole is completed as a 'lateral' that extends parallel with the rock layer containing the substance to be extracted. For example, laterals extend 1,500 to 5,000 feet in the Barnett Shale basin. In contrast, a vertical well only accesses the thickness of the rock layer, typically 50–300 feet. Horizontal drilling also reduces surface disruptions as fewer wells are required. Drilling a wellbore produces rock chips and fine rock particles that may enter cracks and pore space at the wellbore wall, reducing the permeability at and near the wellbore. This reduces flow into the borehole from the surrounding rock formation, and partially seals off the borehole from the surrounding rock. Hydraulic fracturing can be used to restore permeability. Hydraulic fracturing is commonly applied to wells drilled in low permeability reservoir rock. An estimated 90 percent of the natural gas wells in the United States use hydraulic fracturing to produce gas at economic rates.

Fracturing

The fluid injected into the rock is typically a slurry of water, proppants, and chemical additives. Additionally, gels, foams, and compressed gases, including nitrogen, carbon dioxide and air can be injected. Various types of proppant include silica sand, resin-coated sand, and man-made ceramics. These vary depending on the type of permeability or grain strength needed. Sand containing naturally radioactive minerals is sometimes used so that the fracture trace along the wellbore can be measured. Chemical additives are applied to tailor the injected material to the specific geological situation, protect the well, and improve its operation, though the injected fluid is approximately 98-99.5% percent water, varying slightly based on the type of well. The composition of injected fluid is changed during the operation of a well over time, that is initially acid is used to increase permeability, then proppants are used with a gradual increase in their size and/or density, and at the end the well is flushed with water under pressure. Injected fluid is to some degree recovered and is managed by underground injection control, treatment and discharge, recycled and stored in pits or containers while new technology is being developed to better handle wastewater and improve reusability. Although the concentrations of the chemical additives are very low, the recovered fluid may be harmful due in part to hydrocarbons picked up from the formation.

Hydraulic fracturing equipment used in oil and natural gas fields usually consists of a slurry blender, one or more high pressure, high volume fracturing pumps (typically powerful triplex, or quintiplex pumps) and a monitoring unit. Associated equipment includes fracturing tanks, high pressure treating iron, a chemical additive unit (used to accurately monitor chemical addition) low pressure pipes and gauges for flow rate, fluid density, and treating pressure. Fracturing equipment operates over a range of pressures and injection rates, and can reach up to 100 MPa (15,000 psi) and 265 L/s (100 barrels per minute).

Fracture Monitoring

Microseismic monitoring is commonly used to estimate the size and orientation of hydraulically induced fractures. Microseismic activity is measured by placing an array of geophones in a nearby wellbore. By mapping the location of any small seismic events associated with the growing hydraulic fracture, the approximate geometry of the fracture is inferred. Tiltmetre arrays, deployed on the surface or down a well, provide another technology for monitoring the strains produced by hydraulic fracturing.

Emission of gases displaced by hydraulic fracturing into the atmosphere may be detected via atmospheric gas monitoring, and can be quantified directly via the eddy covariance flux measurements.

Completion

Advances in completion technology have led to the emergence of open hole multi-stage fracturing systems. These systems effectively place fractures in specific places in the wellbore, thus increasing the cumulative production in a shorter time frame.

Certain reservoirs such as the Bakken, Barnett Shale, Montney and Haynesville Shale cannot be produced using conventional methods. These formations have begun using high tech completion systems capable of mechanically fracturing at certain intervals.

An alternative to the plug and perf method, multi-stage fracturing systems are capable of stimulating several stages in a single day. Compared to the weeks required by the plug and perf method, cost-effective multi-stage completion systems are quickly becoming sought after technology by oil and natural gas companies.

Terminology

Fracture Gradient : The pressure to fracture the formation at a particular depth divided by the depth. A fracture gradient of 18 kPa/m (0.8 psi/foot) implies that at a depth of 3 km (10,000 feet) a pressure of 54 MPa (8,000 psi) will extend a hydraulic fracture.

ISIP - Initial Shut In Pressure : The pressure measured immediately after injection stops. The ISIP provides a measure of the pressure in the fracture at the wellbore by removing contributions from fluid friction.

Leakoff : Loss of fracturing fluid from the fracture channel into the surrounding permeable rock.

Fracturing fluid : The fluid used during a hydraulic fracture treatment of oil, gas or water wells. The fracturing fluid has two major functions:

1. Open and extend the fracture
2. Transport the proppant along the fracture length.

Proppant : Suspended particles in the fracturing fluid that are used to hold fractures open after a hydraulic fracturing treatment, thus producing a conductive pathway that fluids can easily flow along. Naturally occurring sand grains or artificial ceramic material are common proppants used.

Concise Slang : "Fracing" (sometimes spelled "fracking" primarily in media) is a shortened version of fracturing.

Environmental Concerns

Environmental concerns with hydraulic fracturing include the potential contamination of ground water, risks to air quality, the potential migration of gases and hydraulic fracturing chemicals to the surface, the potential mishandling of waste, and the health effects of these. A 2004 study by the Environmental Protection Agency (EPA) concluded that the injection of hydraulic fracturing fluids into CBM wells posed minimal threat to underground drinking water sources. This study has been criticised for only focusing on the injection of fracking fluids, while ignoring other aspects of the process such as disposal of fluids, and environmental concerns such as water quality, fish kills and acid burns; the study was also concluded before public complaints of contamination started emerging. Largely on the basis of this study, hydraulic fracturing was in 2005 exempted by US Congress from any regulation under the Safe Drinking Water Act.

With the explosive growth of natural gas wells in the US, researcher Valerie Brown predicted in 2007 that "public exposure to the many chemicals involved in energy development is expected to increase over the next few years, with uncertain consequences." As development of natural gas wells in the U.S. since the year 2000 has increased, so too have claims by private well owners of water contamination. This has prompted EPA and others to revisit the topic.

While the EPA recognises the potential for contamination of water by hydraulic fracturing, EPA Administrator Lisa Jackson testified in a Senate Hearing Committee "I'm not aware of any proven case where the fracking process itself has affected water...". There are, however, documented incidents of contamination. In 2006 drilling fluids and methane were detected leaking from the ground near a gas well in Clark, Wyoming; 8 million cubic feet of methane were eventually released, and shallow groundwater was found to be contaminated. In the town of Dimock, Pennsylvania, 13 water wells were contaminated with methane (one of them blew up), and the gas company, Cabot Oil & Gas, had to financially compensate residents and construct a pipeline to bring in clean water; the company continued to deny, however, that any "of the issues in Dimock have anything to do with hydraulic fracturing".

Air Emissions and Pollution

One group of emissions associated with natural gas development and production, are the emissions associated with combustion. These emissions include particulate matter, nitrogen oxides, sulfur oxide, carbon dioxide and carbon monoxide. Another group of emissions that are routinely vented into the atmosphere are those linked with natural gas itself, which is composed of methane, ethane, liquid condensate, and volatile organic compounds (VOCs). The VOCs that are especially impactful on health are benzene, toluene, ethyl benzene, and xylene (referred to as a group, called BTEX). Health effects of exposure to these chemicals include neurological problems, birth defects, and cancer.

VOCs, including BTEX, mixed with nitrogen oxides from combustion and combined with sunlight can lead to ozone formation. Ozone has been shown to impact lung function, increase respiratory illness, and is particularly dangerous to lung development in children. In 2008, measured ambient concentrations in the rural Sublette County Wyoming where ranching and natural gas are the main industries were frequently above the National Ambient Air Quality Standards (NAAQS) of 75ppb and have been recorded as high as 125 ppb.

Groundwater Contamination

A Duke University study published in *Proceedings of the National Academy of Sciences* in 2011 examined methane in groundwater in Pennsylvania and New York states overlying the Marcellus Shale and the Utica Shale. It determined that groundwater tended to contain much higher concentrations of methane near fracking wells, with potential explosion hazard; the methane's isotopic signatures and other geochemical indicators were consistent with it originating in the fracked deep shale formations, rather than any other source. Complaints from a few residents on water quality in a developed natural gas field prompted an EPA groundwater investigation in Wyoming. The EPA reported detections of methane and other chemicals such as phthalates in private water wells.

In Pavillion, Wyoming, the EPA discovered traces of methane and foaming agents in several water wells near a gas rig, though it suggested these chemicals might have come from cleaning products. In DISH, Texas, elevated levels of disulphides, benzene, xylenes and naphthalene have been detected in the air, alongside numerous local complaints of headaches, diarrhea, nosebleeds, dizziness, muscle spasms and other problems. Epidemiological studies that might confirm or rule out any connection between these complaints and fracking are virtually non-existent. Individuals "smell things that don't make them feel well, but we know nothing about cause-and-effect relationships in these cases." In Garfield County, Colorado, another area with a high concentration of drilling rigs, volatile organic compound emissions increased 30% between 2004 and 2006; during the same period there was a rash of health complaints from local residents. The health effects of VOCs are largely unquantified, so any causal relationship is difficult to ascertain; however, some of these chemicals are suspected carcinogens and neurotoxins. Investigators from the Colorado School of Public Health performed a study in Garfield regarding potential adverse health effects, and concluded that residents near gas wells might suffer chemical exposures, accidents from industry operations, and psychological impacts such as depression, anxiety and stress. This study (the only one of its kind to date) was never published, owing to disagreements between community members and the drilling company over the study's methods.

In 2010 the film *Gasland* premiered at the Sundance Film Festival. The filmmaker claims that chemicals including toxins, known carcinogens, and heavy metals polluted the ground water near well sites in Pennsylvania, Wyoming, and Colorado.

A 2011 report by the Massachusetts Institute of Technology addressed groundwater contamination, noting "There has been concern that these fractures can also penetrate shallow freshwater zones and contaminate them with fracturing fluid, but there is no evidence that this is occurring. There is, however, evidence of natural gas migration into freshwater zones in some areas, most likely as a result of substandard well completion practices by a few operators. There are additional environmental challenges in the area of water management, particularly the effective disposal of fracture fluids". This study encourages the use of industry best practices to prevent such events from recurring. Directed by Congress, the U.S. EPA announced in March 2010 that it will examine claims of water pollution related to hydraulic fracturing.

Radioactive Contamination

The New York Times has reported radiation in hydraulic fracturing wastewater released into rivers in Pennsylvania. According to a Times report in February 2011, wastewater at 116 of 179 deep gas wells in Pennsylvania "contained high levels of radiation," but its effect on public drinking water supplies is unknown because water suppliers are required to conduct tests of radiation "only sporadically". The Times stated "never-reported studies" by the EPA and a "confidential study by the drilling industry" concluded that radioactivity in drilling waste cannot be fully diluted in rivers and other waterways. Despite this, as of early 2011 federal and state regulators did not require sewage treatment plants that accept drilling waste (which is mostly water) to test for radioactivity. In Pennsylvania, where the drilling boom began in 2008, most drinking-water intake plants downstream from those sewage treatment plants have not tested for radioactivity since before 2006.

In reply to the Times, the conservative News Corp-owned tabloid New York Post stated that the Pennsylvania Department of

Environmental Protection reported that all samples it took from seven rivers in November and December 2010 "showed levels at or below the normal naturally occurring background levels of radioactivity", and "below the federal drinking water standard for Radium 226 and 228.". However the samples taken by the state at at least one river, (the Monongahela, a source of drinking water for parts of Pittsburgh), were taken upstream from the sewage treatment plants accepting drilling waste water. Furthermore, the New York Times has implicated the DEP in industry-friendly inactivity, such as only making a "request — not a regulation" of gas companies to handle their own flowback waste rather than sending them to public water treatment facilities.

Chemicals Used in Fracturing

Water is by far the largest component of fracking fluids. The initial drilling operation itself may consume from 65,000 gallons to 600,000 gallons of fracking fluids. Over its lifetime an average well will require up to an additional 5 million gallons of water for the initial fracking operation and possible restimulation frac jobs.

Chemical additives used in fracturing fluids typically make up less than 2% by weight of the total fluid. Over the life of a typical well, this may amount to 100,000 gallons of chemical additives. They are biocides, surfactants, adjusting viscosity, and emulsifiers. Many are used in household products such as cosmetics, lotions, soaps, detergents, furniture polishes, floor waxes, and paints. Some are also used in food products. A list of the chemicals that have been used was published in a U.S. House of Representatives Report. Some of the chemicals pose no known health hazards, some others are known carcinogens, some are toxic, some are neurotoxins.

For example: benzene (causes cancer, bone marrow failure), lead (damages the nervous system and causes brain disorders), ethylene glycol (antifreeze, causes death), methanol (highly toxic), boric acid (kidney damage, death), 2-butoxyethanol (causes hemolysis). The 2011 US House of Representatives investigative report on the chemicals used in hydraulic fracturing shows that of the 750 compounds in hydraulic fracturing products "more than 650 of these products contained chemicals that are known or possible human carcinogens, regulated under the Safe Drinking Water Act, or listed as hazardous air pollutants" (12). The report also shows that between 2005 and 2009 279 products (93.6 million gallons-not including water) had at least one component listed as "proprietary" or "trade secret" on their Occupational Safety and Health Administration (OSHA) required Material Safety Data Sheet (MSDS).

The MSDS is a list of chemical components in the products of chemical manufacturers, and according to OSHA, a manufacturer may withhold information designated as "proprietary" from this sheet. When asked to reveal the proprietary components, most companies participating in the investigation were unable to do so, leading the committee to surmise these "companies are injecting fluids containing unknown chemicals about which they may have limited understanding of the potential risks posed to human health and the environment" (12). Third-party laboratories are performing analysis on soil, air, and water near the fracturing sites to measure the level of contamination by each of the chemicals. Each state has a contact person in charge of such regulation.

Another study in 2011, titled "Natural Gas Operations from a Public Health Perspective" and published in *Human and Ecological Risk Assessment: An International Journal* identified 632 chemicals used in natural gas operations. Only 353 of these are well-described in the scientific literature; and of these, more than 75% could affect skin, eyes, respiratory and gastrointestinal systems; roughly 40-50% could affect the brain and nervous, immune and cardiovascular systems and the kidneys; 37% could affect the endocrine system; and 25% were carcinogens or mutagens. The study indicated possible long-term health effects that might not appear immediately. The study recommended full disclosure of all products used, along with extensive air and water monitoring near natural gas operations; it also recommended that fracking's exemption from regulation under the US Safe Drinking Water Act be rescinded.

Earthquakes

A report in the UK concluded that fracking was the likely cause of some small earth tremors that happened during shale gas drilling. In addition the United States Geological Survey (USGS) reports that "Earthquakes induced by human activity have been documented in a few locations" in the United States, Japan, and Canada; "the cause was injection of fluids into deep wells for waste disposal and secondary recovery of oil, and the use of reservoirs for water supplies."

The disposal and injection wells referenced are regulated under the Safe Drinking Water Act and UIC laws and are not wells where hydraulic fracturing is generally performed.

Greenhouse Gas Emissions

The use of natural gas rather than oil or coal is sometimes touted as a way of alleviating global warming: natural gas burns more cleanly, and gas power stations can produce up to 50% less greenhouse gases than coal stations. However, an analysis of the well-to-consumer lifecycle of fracked natural gas concluded that 3.6–7.9% of the methane produced by a well will be leaked into the atmosphere during the well's lifetime.

Because methane is such a potent greenhouse gas, this means that over short timescales, shale gas is actually worse than coal or oil. Methane gradually breaks down in the atmosphere, so that over long periods it is not as problematic as carbon dioxide; yet even if shale gas is burnt in efficient gas power stations, its greenhouse-gas footprint is still worse than coal or oil for timescales of less than fifty years.

Public Relations

The considerable opposition against fracking activities in local townships has led companies to adopt a variety of public relations measures to assuage fears about fracking, including the admitted use of "military tactics to counter drilling opponents".

At a conference where public relations measures were discussed, a senior executive at Anadarko Petroleum was recorded on tape saying, *"Download the US Army / Marine Corps Counterinsurgency Manual, because we are dealing with an insurgency"*, while referring to fracking opponents. Matt Pitzarella, spokesman for the most important fracking company in Pennsylvania, Range Resources, also told other conference attendees that Range employed psychological warfare operations veterans.

According to Pitzarella, the experience learnt in the Middle East has been valuable to Range Resources in Pennsylvania, when dealing with emotionally-charged township meetings and advising townships on zoning and local ordinances dealing with fracking.

Hydraulic Fracturing by Country

Hydraulic fracturing has become a contentious environmental and health issue with France banning the practice and a moratorium in place in New South Wales (Australia), Karoo basin (South Africa), Quebec (Canada), and some of the states of the US.

Australia

On 21 February 2011, the ABC's investigative journalism program Four Corners aired a program showing incidents of gas leaks into the water basin and evidence of contamination by hydraulic fracturing in Chinchilla, Queensland as a result of drilling carried out by a Queensland gas company, QGC.

There is currently a moratorium in place on the practice of hydraulic fracturing in the state of New South Wales.. The moratorium will not affect exploration, drilling core holes and getting core samples. NSW Government's restrictions on hydraulic fracturing apply to new licences only. The NSW Government has banned BTEX chemicals as additives, companies to hold a water licences for extraction of more than three megalitres per year and has banned the use of evaporation ponds.

Canada

Fracking has been in use in Canada at an industrial level since the 1990s. Concerns about fracking began in late July 2011, when the Government of British Columbia gave Talisman Energy a long-term water licence to draw water from the BC Hydro-owned Williston Lake reservoir, for a twenty year term. Fracking has also received criticism in New Brunswick and Nova Scotia, and the Nova Scotia government is currently reviewing the practice, with recommendations expected in March 2012. The practice has been temporarily suspended, in Quebec, pending an environmental review. The Canadian Centre for Policy Alternatives has also expressed concern.

France

Hydraulic fracturing was banned in France in 2011 after public pressure.

New Zealand

In New Zealand, hydraulic fracturing is part of petroleum exploration and extraction on a small scale mainly in Taranaki and concerns have been raised by environmentalists.

South Africa

There is currently a moratorium on hydraulic fracturing in South Africa's Karoo region despite the interest of several energy companies.

United Kingdom

Hydraulic fracturing is currently proceeding in the United Kingdom operated by Cuadrilla and a number of other companies. In Lancashire, operations were suspended after two small earthquakes subsequent to drilling operations.

Several protest groups have started to oppose hydraulic fracturing in the UK such as nationwide groups like Frack Off and local groups like Ribble Estuary Against Fracking and The Vale Says No

United States

Hydraulic fracturing for the purpose of oil, natural gas, and geothermal production was exempted under the Safe Drinking Water Act. This was a result of the signage of the Energy Policy Act of 2005, also known as the Halliburton Loophole because of the involvement of former Halliburton CEO Vice President Dick Cheney's involvement in the passing of this exemption. The result of a 2004 EPA study on coalbed hydraulic fracturing was used to justify the passing of the exemption; however EPA whistleblower Weston Wilson and the Oil and Gas Accountability Project found that critical information was removed from the final report.

Opposers of hydraulic fracturing in the US have focused on this 2005 exemption; however the more primary risk to drinking water is the handling and treatment of wastewater produced by hydraulic fracturing. The EPA and the state authorities do have power "to regulate discharge of produced waters from hydraulic operations" (EPA, 2011) under the Clean Water Act, which is regulated by the National Pollutant Discharge Elimination System (NPDES) permit program "Treatment and Disposal of Wastewater from Shale Gas Extraction | NPDES".

Environmental Protection Agency, n.d. Web. 15 Oct. 2011. Although this waste is regulated, oil and gas exploration and production (E&P) wastes are exempt from Federal Hazardous Waste Regulations under Subtitle C of the Resource Conservation and Recovery Act (RCRA) despite the fact that wastewater from hydraulic fracturing contains toxins such as total dissolved solids (TDS), metals, and radionuclides. About 750 chemicals have been listed as additives for hydraulic fracturing in a report to the US Congress in 2011.

Annulus (Oil Well)

The annulus (yellow area in diagram) of an oil well refers to any void between any piping, tubing or casing and the piping, tubing, or casing immediately surrounding it. It is named after the corresponding geometric concept. The presence of an annulus gives the ability to circulate fluid in the well, provided that excess drill cuttings have not accumulated in the annulus, preventing fluid movement and possibly sticking the pipe in the borehole.

For a new well in the process of being drilled, this would be the void between the drill string and the formation being drilled. An easy way to visualise this would be to stand a straw (purple in diagram) straight up in the centre of a glass of water. All of the water in between the straw and the sides of the glass would be the annulus (yellow area in diagram), with the straw itself representing the drill string and the sides of the glass representing the formation. While drilling, drilling fluid is pumped down the inside of the drill string and pushes the drill cuttings up the annulus to the surface, where the cuttings are removed from the drilling fluid (drilling mud) by the shale shakers.

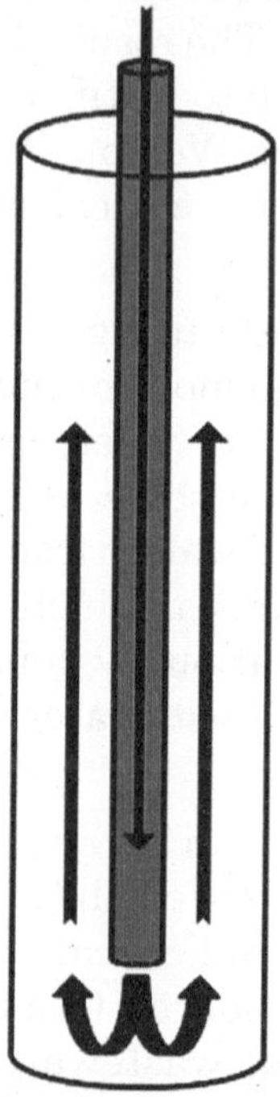

Yellow area represents the Annulus
Arrows indicate the flow of fluid

In a completed well, there may be many annuli. The 'A' annulus is the void between the production tubing and the smallest casing string. The A annulus can serve a number of crucial tasks, including gas lift and well kills. A normal well will also have a 'B' and frequently a 'C' annulus, between the different casing strings. These annuli do not normally have any connection to well bore fluids, but maintaining pressure in them is important in order to ensure integrity of the casing strings. Though all annuli in a completed well are expected to be isolated from the production tubing and each other, connections allowing the flow of fluids between them may sometimes occur, due to either intervention or wear and tear. In these situations, it is said that there is "communication" between them. During coiled tubing interventions, the void between the coil and the production tubing can also be considered an annulus and be used for circulation.

Carbon Dioxide Flooding

Carbon dioxide (CO_2) flooding is a process whereby carbon dioxide is injected into an oil reservoir in order to increase output when extracting oil.

When a reservoir's pressure is depleted through primary and secondary production, Carbon Dioxide flooding can be an ideal tertiary recovery method. It is particularly effective in reservoirs deeper than 2,000 ft., where CO_2 will be in a supercritical state, with API oil gravity greater than 22–25° and remaining oil saturations greater than 20%. It should also be noted that Carbon dioxide flooding is not affected by the lithology of the reservoir area but simply by the reservoir characteristics. Carbon dioxide flooding works on the premise that by injecting CO_2 into the reservoir, the viscosity of any hydrocarbon will be reduced and hence will be easier to sweep to the production well.

If a well has been produced before and has been designated suitable for CO_2 flooding, the first thing to do is to restore the pressure within the reservoir to one suitable for production. This is done by injecting water (with the production well shut off) which will restore pressure within the reservoir to a suitable pressure for CO_2 flooding. Once the reservoir is at this pressure, the next step is to inject the CO_2 into the same injection wells used to restore pressure.

The CO_2 gas is forced into the reservoir and is required to come into contact with the oil. This creates this miscible zone that can be moved easier to the production well. Normally the CO_2 injection is alternated with more water injection and the water acts to sweep the oil towards the production zone. The Weyburn oil field is a famous example where this method is applied in financially interesting conditions.

CO_2 flooding is the second most common tertiary recovery technique and is used in facilities around the world. In the scope of global warming, it is an available method to curb CO_2 emissions.

Coiled Tubing Umbilical

Coiled tubing umbilicals are a type of piping used in and gas wells, which can be used during interventions. They encapsulate multiple coiled tubing and electrical strings into a single string which can be deployed by one conventional unit. These technologies allow for the deployment of complex pumps on coiled tubing. Multiple strings in a casing may maintain a manageable shape for the tubing and eliminating the danger of ovalization.

Drake Well

The Drake Well is a 69.5-foot (21.2 m) deep oil well in Cherrytree Township, Venango County in the U.S. state of Pennsylvania. The well is the centrepiece of the Drake Well Museum located 3 miles (4.8 km) south of Titusville.

Drilled by Edwin Drake in 1859, along the banks of Oil Creek, it is the first commercial oil well in the United States. Drake Well was listed on National Register of Historic Places and designated a National Historic Landmark in 1966. It was designated a Historic Mechanical Engineering Landmark in 1979. The well was designated a National Historic Chemical Landmark in 2009, on the sesquicentennial of the strike.

Location and Geology

The Drake Well is located in Cherrytree Township, Venango County in northwestern Pennsylvania. Situated on the flats 150 feet (46 m) from the east bank of Oil Creek. The site was originally on an artificial island formed by the creek and a mill race. On a floodplain, the well and the museum are protected by an earthen dike.

Most of the oil produced in northwestern Pennsylvania was formed in sandstone reservoir rocks at the boundary between the Mississippian and Devonian rock layers. Over time, the oil migrated toward the surface, became trapped beneath an impervious layer of caprock, and formed a reservoir. The presence of upwards-curving folds in the caprock called anticlines, or sometimes an inversion of an anticline called a syncline, greatly varied the depth of the reservoirs, from around 4,000 feet (1,200 m) to just beneath the surface.

History

Petroleum found along Oil Creek was known to Native Americans for hundreds of years through natural seeps. Europeans became aware of the existence of petroleum in the 1600s. At the time, this "mineral-oil" was used primarily for medicinal purposes and was reputed to cure many ailments, including rheumatism and arthritis. Around 1848, Samuel Kier realised the potential of the medicinal oil as an illuminant. Kier distilled the oil to make it more suitable in lamps by removing the odor and impurities that created soot when burned. A sample of oil was brought to Dartmouth College by Francis B. Brewer from the Watson, Brewer and Company Farm on Oil Creek around 1853. The sample was acquired by George Bissell who, along with Jonathan G. Eveleth purchased the farm for $5,000. Bissell and Eveleth took another sample of oil to Benjamin Silliman at Yale University in 1855 for further investigation. Silliman's report confirmed the quality of the petroleum and described the distillation processes needed to produce kerosene. The Pennsylvania Rock Oil Company was incorporated and the farm transferred to the company.

Construction and Operation

Edwin Drake, a former conductor for the New York and New Haven Railroad, invested $200, his entire savings, into the Pennsylvania Rock Oil Company. Drake became more involved in the company and travelled to Titusville, Pennsylvania and the Brewer and Watson Farm in December 1857. His report prompted Bissell and Eveleth to organise the Seneca Oil Company in Connecticut in March 1858 and to place Drake in charge of producing petroleum. Because Drake decided that drilling in the manner of salt wells would yield more petroleum than conventional digging, he hired William A. Smith, a Tarentum blacksmith and salt-well driller, to aid in the endeavour. An engine house and derrick were constructed, and Drake purchased a 6-horsepower (4.5 kW), horizontal steam engine. The steam engine was used to ram the drill through the soil until it reached bedrock 32 feet (9.8 m) down. After it was found that groundwater would cause the walls of the hole to collapse, Drake acquired 50 feet (15 m) of cast iron pipe to stabilise the hole. After reaching bedrock, Drake and Smith were able to drill at a rate of 3 feet (0.91 m) per day. Drake's colleagues back in Connecticut gave up on finding any oil by April 1859 and after spending $2,500, Drake took out a $500 loan to keep the operation going. The drill reached its maximum depth of 69.5 feet (21.2 m) on August 27, 1859. Smith visited the well the next day and found oil visible on top of the water 5 inches (130 mm) from the top of the well. The original structures at the well caught fire in October 1859 and were rebuilt by Drake a month later. The well produced 12 to 20 barrels (1.9 to 3.2 m^3) a day, but, after the price of plummeted from the resulting oil boom, it was never profitable. The well stopped producing in 1861 and the Seneca Oil Company sold the property in 1864. The derrick was moved in 1876 to the Centennial Exposition in Philadelphia.

Preservation

The well remained abandoned until 1889, when David Emery of Titusville bought the site, erected a derrick and cleaned out the well. Emery was able to obtain a small quantity of petroleum from the well and attempted to sell it as souvenirs to raise funds "to perpetuate the site", but died before he was able to do so. His widow donated the 1 acre (0.40 ha) that included the well to the Canadohta Chapter of the

Daughters of the American Revolution in 1913. The chapter erected a limestone boulder with a bronze plaque at the well in 1914 to commemorate the site. In 1931, the American Petroleum Institute donated $60,000 for the creation of a museum and library, as well as a dike to protect Drake Well from flooding by Oil Creek. The Institute stipulated that when the Commonwealth of Pennsylvania took ownership of the site during the Diamond Jubilee of Drake Well in 1934, it was to be made a state park. The Drake Well State Park remained under the control of the Department of Forestry and Waters, the precursor to the Pennsylvania Department of Conservation and Natural Resources, until 1943 when it was transferred along with the Cornwall Iron Furnace in Lebanon County to the Pennsylvania Historical Commission.

Since at least the late-1890s, the only artifacts remaining from the original well were the drilling tools and drive pipe, much to the disappointment of visitors to Drake Well. In 1945, the Pennsylvania General Assembly appropriated $185,000 for the construction of a replica derrick and engine house, including pumping equipment. The "board-for-board replica" was duplicated from photographs of the well taken by John A. Mather from the 1860s. Drake Well was listed on National Register of Historic Places and designated a National Historic Landmark on November 13, 1966 by the National Park Service. It was designated a Historic Mechanical Engineering Landmark by the American Society of Mechanical Engineers in October 1979. Authentic reproductions of the steam engine and boiler were purchased in Erie and installed in 1986. The American Chemical Society designated the Drake Well a National Historic Chemical Landmark on August 27, 2009, the 150th anniversary of the strike

Museum

The Drake Well Museum encompasses 22 acres (8.9 ha) of land that surrounds the well. The museum was accredited by the American Association of Museums in 1983 and reaccredited in 1995. The museum has a station on the Oil Creek and Titusville Railroad.

Bibliography

Arthur W. McCray and Frank W. Cole: *Oil Well Drilling Technology*, New India Pub, Delhi, 2005.

Black, Brian: *Petrolia: The Landscape of America's First Oil Boom.* Baltimore: Johns Hopkins University Press, 2000.

Brantly, J.E.: History of Oil Well Drilling, Gulf Publishing Co., Houston, 1971.

Chen and Lui: *Handbook of Structural Engineering,* CRC Press, NY, 2005.

Chowdhary, L.R. : *Petroleum Geology of the Cambay Basin, Gujarat, India,* Indian Petroleum, Delhi, 2004.

Chris Woodford: *Cool Stuff and How It Works.* DK Publishing, Inc., New York, 2005.

Cording, E.J., and Deere, D.U.: *Rock Tunnel Supports and Field Measurement,* AIME, Chicago, 1972.

Cowhey, Peter F.: *The Problems of Plenty: Energy Policy and International Politics.* Berkeley: Uni-versity of California Press, 1985.

Critchfield, J. W.: *Tunneling in Gassy Ground,* Proceedings, RETC, AIME, New York, 1985.

DeLaisi, Francis: *Oil, Its Influence on Politics.* Labor Publishing,1923.

Dowding, C. H.: *Blast Vibration Monitoring and Control,* Prentice Hall, New York, 1985.

Engler, Robert: *The Politics of Oil: A Study of Private Power and Democratic Directions.* New York, Macmillan, 1961.

Fischer, Louis: *Oil Imperialism: The International Struggle for Petroleum.* New York: International Publishers, 1926.

Frankel, Paul: *Mattei: Oil and Power Politics.* New York and Washington: Praeger, 1966.

Gramling, Robert: *Oil on the Edge: Offshore Development, Conflict, Gridlock.* Albany: State University of New York Press, 1996.

Hansmire, W. and Monsees, J.E.: *Evolving Practices for Tunnel Contracts in the United States,* Elsevier, London, 2006,

Hartshorn, J. E.: *Oil Companies and Governments: An Account of the International Oil Industry in Its Political Environment.* London: Faber and Faber, 1962.

Jentleson, Bruce: *Pipeline Politics: The Complex Political Economy of East-West Energy Trade.* lthaca: Cornell University Press, 1986.

Klare, Michael: *Blood and Oil: The Dangers and Consequences of America's Growing Petroleum Diplomacy.* Metropolitan Books/ Henry Holt, 2004.

Kuesel, T.: *Immersed Tube Tunnels: Concept Design and Construction,* Boston Society of Civil Engineers, Spring, 1986.

Lieber, Robert J.: *Oil and the Middle East War: Europe in the Energy Crisis.* Cambridge: Center for International Affairs, 1976.

Olien, Diana Davids: *Oil in Texas: The Gusher Age, 1895–1945.* Austin: University of Texas Press, 2002.

Paretsky, Sara: *Tunnel Vision,* Dell Publishing, New York, 1994.

Phien-Wej, N., and EJ. Cording: *Sheared Shale Response to Deep TBM Excavation,* Elsevier, New York, 1991.

Philip, George: *Oil and Politics in Latin America: Nationalist Movements and State Companies.* Cambridge University Press, Cambridge, 1982.

Proctor, R.Y. and T. L. White: *Earth Drilling with Steel Supports,* Commercial Shearing, Inc., Youngstown, OH, 1977.

Ripley, B. D: *Design of Pressure Tunnels and Shafts,* J. A. Hudson, 1993.

Roark, R.J., and Yomg,W.C.: *Formulas for Stvess and Strain,* McGraw-Hill, New York, 1975.

Satyanarayana, D. : *Petroleum Geochemistry,* Daya Publishing, Delhi, 2011.

Schneider, Steven A.: *The Oil Price Revolution.* Baltimore, 1983.

Stack, B.: *Handbook of Mining and Drilling Machinery,* Wiley, New York, 1982.

Teng, W.C.: *Foundation Design of Drilling,* Prentice-Hall, Inc., Englewood, NJ, 1962.

Tiwari, S.K. : *Fuel Minerals and Other Energy Resources,* Atlantic Pub, Delhi, 2010.

Tschebotarioff, G.P.: *Retaining Structures, Foundation Engineering,* McGraw-Hill, New York, 1962.

William H. Rae, Jr.: *Low-Speed Wind Tunnel Testing,* John Wiley & Sons, Inc., New York, 1999.

Index

❑❑❑